Spring Security 实战

陈木鑫◎编著

电子工业出版社
Publishing House of Electronics Industry
北京·BEIJING

内 容 简 介

Spring Security 是一个强大且高度可定制的安全框架,致力于为 Java 应用提供身份认证和授权。

本书通过 4 部分内容由浅入深地介绍了 Spring Security 的方方面面。第 1 部分主要讲解 Spring Security 的基本配置;第 2 部分剖析 Web 项目可能遇到的安全问题,并讲解如何使用 Spring Security 进行有效防护;第 3 部分详细介绍 OAuth,并使用 Spring Social 整合 Spring Security,实现 QQ 快捷登录;第 4 部分重点介绍 Spring Security OAuth 框架,剖析 Spring Security OAuth 的部分核心源码。

本书适合有一定 Java 基础的读者,以及希望在项目中应用 Spring Security 的开发人员阅读。

未经许可,不得以任何方式复制或抄袭本书之部分或全部内容。
版权所有,侵权必究。

图书在版编目(CIP)数据

Spring Security 实战 / 陈木鑫编著. —北京:电子工业出版社,2019.8
ISBN 978-7-121-37143-1

Ⅰ. ①S… Ⅱ. ①陈… Ⅲ. ①JAVA 语言—程序设计 Ⅳ. ①TP312.8

中国版本图书馆 CIP 数据核字(2019)第 152390 号

责任编辑:安　娜
印　　刷:北京捷迅佳彩印刷有限公司
装　　订:北京捷迅佳彩印刷有限公司
出版发行:电子工业出版社
　　　　　北京市海淀区万寿路 173 信箱　邮编:100036
开　　本:787×980　1/16　印张:17.75　字数:348 千字
版　　次:2019 年 8 月第 1 版
印　　次:2022 年 3 月第 6 次印刷
定　　价:79.00 元

凡所购买电子工业出版社图书有缺损问题,请向购买书店调换。若书店售缺,请与本社发行部联系,联系及邮购电话:(010)88254888,88258888。
质量投诉请发邮件至 zlts@phei.com.cn,盗版侵权举报请发邮件至 dbqq@phei.com.cn。
本书咨询联系方式:010-51260888-819,faq@phei.com.cn。

前　言

Spring Security 的前身是 Acegi Security，在被收纳为 Spring 子项目后正式更名为 Spring Security。在笔者成书时，Spring Security 已经升级到 5.1.3.RELEASE 版本，不仅新增了原生 OAuth 框架，还支持更加现代化的密码加密方式。可以预见，在 Java 应用安全领域，Spring Security 会成为首先被推崇的安全解决方案。

虽然 Spring Security 有强大的功能，但它同时也有很高的学习成本。它囊括了身份认证的各种应用场景以及 Web 安全的大量知识，仅官方参考手册就有数十万字，并且还省略了诸多实现细节。许多开发人员在面对这样的"庞然大物"时无从入手，更因为对其不够了解而在实际项目中不敢轻易采用。

本书由浅入深、抽丝剥茧地讲解了 Spring Security 的典型应用场景，另外，还分析了部分核心源码，以及许多开发语言之外的安全知识。通过本书，读者不仅可以学习如何应用 Spring Security，还可以学习借鉴它的实现思路，以将这种实现思路应用到其他开发场景中。

本书读者

本书主要面向有一定 Java 基础的读者，以及希望在实际项目中应用 Spring Security 的开发人员。

本书内容

本书共分为 4 个部分。

第 1 部分（第 1 章至第 3 章）主要介绍 Spring Security 的基本配置，包括默认配置、简单表单认证，以及基于数据库模型的认证与授权。

第 2 部分（第 4 章至第 11 章）主要介绍各种定制化的配置场景，剖析 Web 项目可能遇到的安全问题，并讲解如何使用 Spring Security 进行有效防护，部分章节还配备了详细的源码导读。

第 3 部分（第 13 章）将登录用户的数据来源从系统内转移到社交平台，详细介绍了 OAuth，并使用 Spring Social 整合 Spring Security，实现 QQ 快捷登录，满足一般性的项目需求。

第 4 部分（第 14 章）带领读者认识 Spring Security OAuth 框架，并基于该框架完整实现了 OAuth 客户端、OAuth 授权服务器以及 OAuth 资源服务器三种角色。除此之外，还简单剖析了 Spring Security OAuth 的部分核心源码，以帮助读者更好地理解 OAuth 框架。

致谢

首先感谢赵召同学（Andy）对第 4 部分的贡献。笔者在 Gitter 发言表明会写一本关于 Spring Security 的中文书后，赵召同学找到了我，并希望与我一起写作，但由于这本书实际上已基本成型，所以赵召同学贡献了第 4 部分，使得本书内容更加充实，再次感谢赵召同学的贡献。

其次也非常感谢博文视点公司的安娜编辑以及身边的朋友给予的鼓励。从下定决心编写这本书开始，笔者其实经历了非常多的折磨，不管是思路的枯竭还是耐心的消磨，都致使笔者几次三番萌生退意，但最终还是在不断的鼓励声中坚持了下来，成功为国内希望学习 Spring Security 的朋友奉上了一本中文版的教程，这份收获也应当属于他们。

陈木鑫

读者服务

轻松注册成为博文视点社区用户（www.broadview.com.cn），扫码直达本书页面。

- ◎ **提交勘误**：您对书中内容的修改意见可在【提交勘误】处提交，若被采纳，将获赠博文视点社区积分（在您购买电子书时，积分可用来抵扣相应金额）。
- ◎ **与读者交流**：在页面下方【读者评论】处留下您的疑问或观点，与其他读者一同学习交流。

页面入口：http://www.broadview.com.cn/37143

目 录

第 1 部分

第 1 章 初识 Spring Security 2
    1.1 Spring Security 简介 2
    1.2 创建一个简单的 Spring Security 项目 4

第 2 章 表单认证 10
    2.1 默认表单认证 10
    2.2 自定义表单登录页 13

第 3 章 认证与授权 19
    3.1 默认数据库模型的认证与授权 19
        3.1.1 资源准备 19
        3.1.2 资源授权的配置 20
        3.1.3 基于内存的多用户支持 22
        3.1.4 基于默认数据库模型的认证与授权 23
    3.2 自定义数据库模型的认证与授权 28
        3.2.1 实现 UserDetails 28
        3.2.2 实现 UserDetailsService 32

第 2 部分

第 4 章 实现图形验证码 36
    4.1 使用过滤器实现图形验证码 36

4.1.1 自定义过滤器 36
4.1.2 图形验证码过滤器 39
4.2 使用自定义认证实现图形验证码 44
4.2.1 认识 AuthenticationProvider 44
4.2.2 自定义 AuthenticationProvider 47
4.2.3 实现图形验证码的 AuthenticationProvider 53

第 5 章 自动登录和注销登录 59
5.1 为什么需要自动登录 59
5.2 实现自动登录 60
5.3 注销登录 70

第 6 章 会话管理 75
6.1 理解会话 75
6.2 防御会话固定攻击 76
6.3 会话过期 78
6.4 会话并发控制 79
6.5 集群会话的缺陷 93
6.6 集群会话的解决方案 94
6.7 整合 Spring Session 解决集群会话问题 95

第 7 章 密码加密 98
7.1 密码安全的重要性 98
7.2 密码加密的演进 98
7.3 Spring Security 的密码加密机制 102

第 8 章 跨域与 CORS 108
8.1 认识跨域 108
8.2 实现跨域之 JSONP 109
8.3 实现跨域之 CORS 111
8.4 启用 Spring Security 的 CORS 支持 113

第 9 章　跨域请求伪造的防护 116
9.1　CSRF 的攻击过程 116
9.2　CSRF 的防御手段 117
9.3　使用 Spring Security 防御 CSRF 攻击 118

第 10 章　单点登录与 CAS 128
10.1　单点登录 128
10.2　认识 CAS 131
10.3　搭建 CAS Server 133
10.4　用 Spring Security 实现 CAS Client 140

第 11 章　HTTP 认证 146
11.1　HTTP 基本认证 146
11.2　HTTP 摘要认证 147
11.2.1　认识 HTTP 摘要认证 147
11.2.2　Spring Security 对 HTTP 摘要认证的集成支持 148
11.2.3　编码实现 150

第 12 章　@EnableWebSecurity 与过滤器链机制 153
12.1　@EnableWebSecurity 153
12.2　WebSecurityConfiguration 154

第 3 部分

第 13 章　用 Spring Social 实现 OAuth 对接 163
13.1　OAuth 简介 163
13.1.1　什么是 OAuth 163
13.1.2　OAuth 的运行流程 165
13.2　QQ 互联对接准备 169
13.2.1　申请 QQ 互联应用 170
13.2.2　QQ 互联指南 171

13.2.3 回调域名准备 175
13.3 实现 QQ 快捷登录 177
13.3.1 引入 Spring Social 177
13.3.2 新增 OAuth 服务支持的流程 179
13.3.3 编码实现 180
13.4 与 Spring Security 整合 192
13.5 Spring Social 源码分析 194
13.5.1 SocialAuthenticationFilter 194
13.5.2 OAuth2AuthenticationService 196
13.5.3 OAuth2Connection 197
13.5.4 OAuth2Template 198
13.5.5 SocialAuthenticationProvider 199
13.5.6 JdbcUsersConnectionRepository 200
13.6 配置相关 201

第 4 部分

第 14 章 用 Spring Security OAuth 实现 OAuth 对接 206

14.1 实现 GitHub 快捷登录 207
14.2 用 Spring Security OAuth 实现 QQ 快捷登录 210
14.2.1 OAuth 功能扩展流程 210
14.2.2 编码实现 212
14.2.3 自定义 login.html 和 index.html 223
14.2.4 自定义 Controller 映射 224
14.2.5 启用自定义登录页 225
14.3 OAuth Client 功能核心源码分析 226
14.3.1 OAuth2AuthorizationRequestRedirectFilter 227
14.3.2 OAuth2LoginAuthenticationFilter 228
14.3.3 DefaultLoginPageGeneratingFilter 230
14.3.4 OAuth2LoginAuthenticationProvider 231
14.4 Spring Security OAuth 授权服务器 232

  14.4.1　功能概述 233
  14.4.2　依赖包说明 233
  14.4.3　编码实现 234
14.5　OAuth 授权服务器功能扩展和自定义配置 236
  14.5.1　自定义配置的授权服务器 237
  14.5.2　编写 OAuth 客户端 247
  14.5.3　使用 JDBC 存储 OAuth 客户端信息 248
  14.5.4　使用 JDBC 存储 token 254
  14.5.5　其他功能配置 255
14.6　实现 OAuth 资源服务器 255
  14.6.1　依托于授权服务器的资源服务器 256
  14.6.2　独立的资源服务器 258
14.7　Spring Security OAuth 核心源码分析 263
  14.7.1　授权服务器核心源码分析 264
  14.7.2　资源服务器核心源码分析 271

第 1 部分

第1章
初识Spring Security

本书所有的示例都基于 Intellij IDEA 创建的 Spring Boot 项目，因此读者需要具备一定的 Spring 相关知识。

1.1 Spring Security 简介

Spring Security 的前身是 Acegi Security，在被收纳为 Spring 子项目后正式更名为 Spring Security。

在笔者成书时，Spring Security 已经升级到 5.1.3.RELEASE 版本，加入了原生 OAuth2.0 框架，支持更加现代化的密码加密方式。可以预见，在 Java 应用安全领域，Spring Security 会成为被首先推崇的解决方案，就像我们看到服务器就会联想到 Linux 一样顺理成章。

应用程序的安全性通常体现在两个方面：认证和授权。

认证是确认某主体在某系统中是否合法、可用的过程。这里的主体既可以是登录系统的用户，也可以是接入的设备或者其他系统。

授权是指当主体通过认证之后，是否允许其执行某项操作的过程。

这些概念并非 Spring Security 独有，而是应用安全的基本关注点。Spring Security 可以帮助我们更便捷地完成认证和授权。

Spring Security 支持广泛的认证技术，这些认证技术大多由第三方或相关标准组织开发。Spring Security 已经集成的认证技术如下：

◎ HTTP BASIC authentication headers：一个基于 IETF RFC 的标准。

- HTTP Digest authentication headers：一个基于 IETF RFC 的标准。
- HTTP X.509 client certificate exchange：一个基于 IETF RFC 的标准。
- LDAP：一种常见的跨平台身份验证方式。
- Form-based authentication：用于简单的用户界面需求。
- OpenID authentication：一种去中心化的身份认证方式。
- Authentication based on pre-established request headers：类似于 Computer Associates SiteMinder，一种用户身份验证及授权的集中式安全基础方案。
- Jasig Central Authentication Service：单点登录方案。
- Transparent authentication context propagation for Remote Method Invocation (RMI) and HttpInvoker：一个 Spring 远程调用协议。
- Automatic "remember-me" authentication：允许在指定到期时间前自行重新登录系统。
- Anonymous authentication：允许匿名用户使用特定的身份安全访问资源。
- Run-as authentication：允许在一个会话中变换用户身份的机制。
- Java Authentication and Authorization Service：JAAS，Java 验证和授权 API。
- Java EE container authentication：允许系统继续使用容器管理这种身份验证方式。
- Kerberos：一种使用对称密钥机制，允许客户端与服务器相互确认身份的认证协议。

除此之外，Spring Security 还引入了一些第三方包，用于支持更多的认证技术，如 JOSSO 等。如果所有这些技术都无法满足需求，则 Spring Security 允许我们编写自己的认证技术。因此，在绝大部分情况下，当我们有 Java 应用安全方面的需求时，选择 Spring Security 往往是正确而有效的。

Internet 工程任务组（Internet Engineering Task Force，IETF）是推动 Internet 标准规范制定的最主要的组织。请求注解（Request For Comments，RFC）包含大多数关于 Internet 的重要文字资料，被称为"网络知识圣经"。

在授权上，Spring Security 不仅支持基于 URL 对 Web 的请求授权，还支持方法访问授权、对象访问授权等，基本涵盖常见的大部分授权场景。

很多时候，一个系统的安全性完全取决于系统开发人员的安全意识。例如，在我们从未听过 SQL 注入时，如何意识到要对 SQL 注入做防护？关于 Web 系统安全的攻击方式非常多，诸如 XSS、CSRF 等，未来还会暴露出更多的攻击方式，我们只有在充分了解其攻击原理后，才能提出完善而有效的防护策略。在笔者看来，学习 Spring Security 并非局限于降低 Java 应用的安全开发成本，通过 Spring Security 了解常见的安全攻击手段以及对应的防护方法也尤为重要，这些是脱离具体开发语言而存在的。

1.2 创建一个简单的 Spring Security 项目

本节创建一个简单的 Spring Security 项目，带领大家初步领略 Spring Security 带来的便利。下面我们就完整地"走"一遍创建项目的流程。

通过 Intellij IDEA 创建 Spring Boot 项目的方式有许多种，其中最简单的方式就是使用 Spring Initializr 工具，省略了在 https://start.spring.io 中生成并导入 Intellij IDEA 的过程。Eclipse 也提供了一个可以实现类似功能的插件：STS（Spring Tool Suite），感兴趣的读者可以自行了解。此处我们单击"Next"按钮进入下一个页面，如图 1-1 所示。

图 1-1

在图 1-2 所示页面中，除可以设置项目的 Group、Artifact 这些基本信息外，还有其他几个配置可选。例如，对于 Type 属性，可以选择 Maven Project 或者 Gradle Project 作为项目管理工具；对于 Language 属性，可以选择使用 Java、Kotlin 或 Groovy 作为开发语言。

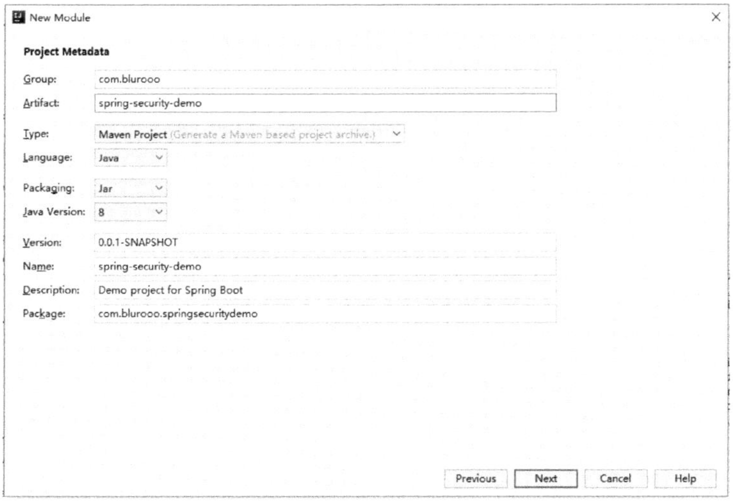

图 1-2

Spring Initializr 将根据我们的选择自动构建项目骨架，选好之后单击"Next"按钮进入下一个页面，如图 1-3 所示。

图 1-3

Spring Initializr 允许我们提前选定一些常用的项目依赖，此处我们选择 Security 作为构建 Spring Security 项目的最小依赖，选择 Web 作为 Spring Boot 构建 Web 应用的核心依赖。

当项目创建完成后，可以得到如图 1-4 所示的目录结构。

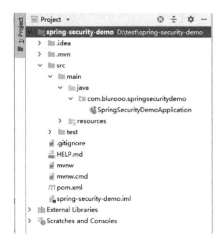

图 1-4

打开 pom.xml 文件，看看 Spring Initializr 引入了哪些依赖。

```xml
<dependencies>
    <dependency>
        <groupId>org.springframework.boot</groupId>
        <artifactId>spring-boot-starter</artifactId>
        <version>2.1.9.RELEASE</version>
        <scope>compile</scope>
    </dependency>
    <dependency>
        <groupId>org.springframework</groupId>
        <artifactId>spring-aop</artifactId>
        <version>5.1.10.RELEASE</version>
        <scope>compile</scope>
    </dependency>
    <dependency>
        <groupId>org.springframework.security</groupId>
        <artifactId>spring-security-config</artifactId>
        <version>5.1.6.RELEASE</version>
        <scope>compile</scope>
    </dependency>
    <dependency>
        <groupId>org.springframework.security</groupId>
        <artifactId>spring-security-web</artifactId>
        <version>5.1.6.RELEASE</version>
        <scope>compile</scope>
    </dependency>
</dependencies>
```

从代码中可以看到，在选择 Security 之后，Spring Initializr 自动引入 spring-security-web 和 spring-security-config 两个核心模块，这正是官方建议引入的 Spring Security 最小依赖。当需要引入更多的 Spring Security 特征时，再编辑 pom.xml 文件即可。如果不通过 Spring Initializr 添加 Spring Security 的相关依赖，则手动将依赖信息添加到 pom.xml 文件也是可以的。

```xml
<dependency>
        <groupId>org.springframework.boot</groupId>
        <artifactId>spring-boot-starter-security</artifactId>
</dependency>
```

或者跳过 Spring Boot 自动配置选项，直接添加 Spring Security 相关模块。

```xml
<dependencies>
    <!-- ... 其它依赖 ... -->
    <dependency>
        <groupId>org.springframework.security</groupId>
        <artifactId>spring-security-web</artifactId>
        <version>5.1.6.RELEASE</version>
    </dependency>
    <dependency>
        <groupId>org.springframework.security</groupId>
        <artifactId>spring-security-config</artifactId>
        <version>5.1.6.RELEASE</version>
    </dependency>
</dependencies>
```

同理，通过 Gradle 管理的项目只需引入 spring-security-web 和 spring-security-config 两个核心模块。

```
dependencies {
        compile 'org.springframework.security:spring-security-web:5.1.6.RELEASE'
        compile 'org.springframework.security:spring-security-config:5.1.6.RELEASE'
}
```

下面打开程序的入口类 SpringSecurityDemoApplication，声明一个 hello 路由。

```java
@RestController
@SpringBootApplication
public class SpringSecurityDemoApplication {

        @GetMapping("/")
```

```
        public String hello() {
                return "hello, spring security";
        }

        public static void main(String[] args) {
                SpringApplication.run(SpringSecurityDemoApplication.class,
args);
        }
}
```

首先选中 SpringSecurityDemoApplication 类并单击运行项（如果没有出现运行项，则可能是因为 Maven 没有初始化。选中 pom.xml，单击右键，从右键快捷菜单中选择 Add as maven build file 选项，重新构建即可）。在默认情况下，项目将成功启动并监听 8080 端口。如果运行失败，则可能是因为 8080 端口被占用。

接着打开浏览器访问 localhost:8080，将出现一个表单登录页面，如图 1-5 所示。

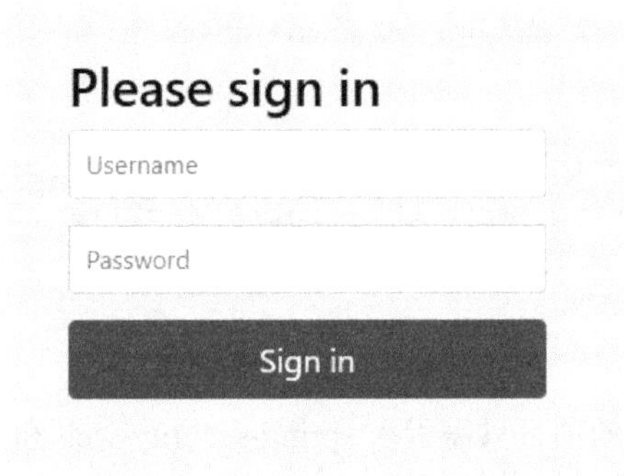

图 1-5

在引入 Spring Security 项目之后，虽然没有进行任何相关的配置或编码，但 Spring Security 有一个默认的运行状态（在 WebSecurityConfigurerAdapter 这个抽象类中实现，由 Spring Boot 自动配置生效），要求在经过 HTTP 表单认证后才能访问对应的 URL 资源，其默认使用的用户名为 user，密码则是动态生成并打印到控制台的一串随机码。翻看控制台的打印信息，可以看到如图 1-6 所示的输出。

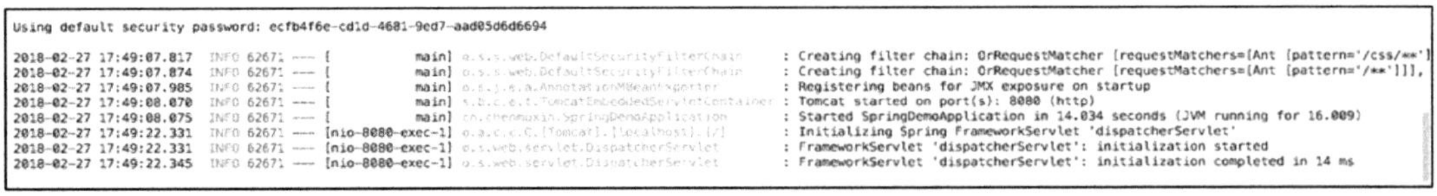

```
Using default security password: ecfb4f6e-cd1d-4681-9ed7-aad05d6d6694

2018-02-27 17:49:07.817  INFO 62671 --- [           main] o.s.s.web.DefaultSecurityFilterChain     : Creating filter chain: OrRequestMatcher [requestMatchers=[Ant [pattern='/css/**']
2018-02-27 17:49:07.874  INFO 62671 --- [           main] o.s.s.web.DefaultSecurityFilterChain     : Creating filter chain: OrRequestMatcher [requestMatchers=[Ant [pattern='/**']]],
2018-02-27 17:49:07.985  INFO 62671 --- [           main] o.s.j.e.a.AnnotationMBeanExporter        : Registering beans for JMX exposure on startup
2018-02-27 17:49:08.070  INFO 62671 --- [           main] s.b.c.e.t.TomcatEmbeddedServletContainer : Tomcat started on port(s): 8080 (http)
2018-02-27 17:49:08.075  INFO 62671 --- [           main] cn.chenmuxin.SpringDemoApplication       : Started SpringDemoApplication in 14.034 seconds (JVM running for 16.009)
2018-02-27 17:49:22.331  INFO 62671 --- [nio-8080-exec-1] o.a.c.c.C.[Tomcat].[localhost].[/]       : Initializing Spring FrameworkServlet 'dispatcherServlet'
2018-02-27 17:49:22.331  INFO 62671 --- [nio-8080-exec-1] o.s.web.servlet.DispatcherServlet        : FrameworkServlet 'dispatcherServlet': initialization started
2018-02-27 17:49:22.345  INFO 62671 --- [nio-8080-exec-1] o.s.web.servlet.DispatcherServlet        : FrameworkServlet 'dispatcherServlet': initialization completed in 14 ms
```

图 1-6

输入用户名和密码后，单击"登录"按钮即可成功访问 hello 页面，如图 1-7 所示。

图 1-7

当然，用户名和密码都是可以配置的，最常见的方式就是在 resources 下的配置文件中修改，如图 1-8 所示。

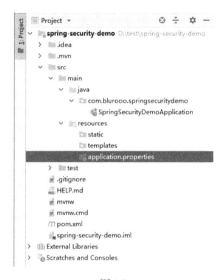

图 1-8

打开 application.properties，输入以下配置信息：

```
spring.security.user.name=user
spring.security.user.password=123
```

重新启动程序，发现控制台不再打印默认密码串了，此时使用我们自定义的用户名和密码即可登录。

在 4.x 的版本中，Spring Boot 默认为 Spring Security 配置的登录方式是 HTTP 基本认证，即用户名和密码在弹框中完成。事实上，绝大部分 Web 应用都不会选择 HTTP 基本认证这种方式，除安全性差、无法携带 cookie 等因素外，灵活性不足也是它的一个主要缺点。通常大家更愿意选择表单认证，所以在后来的 Spring Boot 版本中便把 HTTP 基本认证给移除了，取而代之的是 Spring Security 的默认配置策略。

第 2 章 表单认证

在第 1 章中,我们初步引入了 Spring Security,并使用其默认生效的 HTTP 表单认证来保护 URL 资源,本章我们将继续学习表单认证的更多内容。

2.1 默认表单认证

首先,新建一个 configuration 包用于存放通用配置;然后,新建一个 WebSecurityConfig 类,使其继承 WebSecurityConfigurerAdapter,如图 2-1 所示。

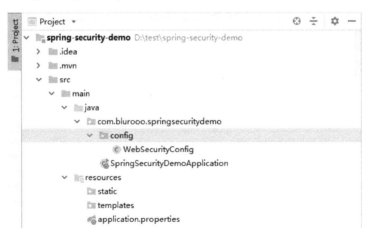

图 2-1

在给 WebSecutiryConfig 类中加上@EnableWebSecurity 注解后,便会自动被 Spring 发现并注册(查看@EnableWebSecurity 即可看到@Configuration 注解已经存在,所以此处不需要额外添加)。

```
@EnableWebSecurity
public class WebSecurityConfig extends WebSecurityConfigurerAdapter {

}
```

接着查看 WebSecurityConfigurerAdapter 类对 configure(HttpSecurity http)的定义。

```
protected void configure(HttpSecurity http) throws Exception {
        logger.debug("Using default configure(HttpSecurity). If subclassed this will potentially override subclass configure(HttpSecurity).");

        http
                .authorizeRequests()
                        .anyRequest().authenticated()
                        .and()
                .formLogin().and()
                .httpBasic();
}
```

可以看到 WebSecurityConfigurerAdapter 已经默认声明了一些安全特性：

◎ 验证所有请求。
◎ 允许用户使用表单登录进行身份验证（Spring Security 提供了一个简单的表单登录页面）。
◎ 允许用户使用 HTTP 基本认证。

现在重启服务，应用新的安全配置。可以预见，在下次访问 localhost:8080 时，系统会要求我们进行表单认证。

如果是 Spring Security 4.x 版本，由于 Spring Boot 1.x 默认为我们配置了 HTTP 基本认证，恰好我们也在前面一章中顺利通过认证的话，结果就会很意外，无论怎么刷新，都无法看到表单登录页，为什么呢？我们可以从浏览器的开发者工具中找到蛛丝马迹（在 Chrome 浏览器中按快捷键 F12 调出开发者工具），如图 2-2 所示。

可以清楚地看到，浏览器发出的请求头中自动携带 Authorization 属性，由于 Spring Security 的配置刚好同时支持 HTTP 基本认证，所以并不需要在表单中重新登录。

这实际上属于浏览器的默认行为，只要在 HTTP 基本认证中成功认证过，便会自动记住一段时间，也就是说，可以跳过登录直接访问系统资源。那么如何避免这种情况的发生呢？在 IE 浏览器中，可以在控制台执行 document.execCommand("ClearAuthenticationCache")语句清除

HTTP 基本认证缓存，但这种方法在 Chrome 浏览器中并不适用。建议调试时直接使用浏览器的无痕模式，简单方便，可以避免很多缓存问题（在 Windows 下的 Chrome 浏览器中，可用组合键 Ctrl+Shift+n 打开浏览器的无痕模式；在 mac 下的 Chrome 浏览器中，可用组合键 Command+Shift+n 打开浏览器的无痕模式）。

图 2-2

经过一些小插曲，我们终于成功进入表单登录页，如图 2-3 所示。

图 2-3

从图 2-3 中的地址栏可以发现，我们访问的地址自动跳转到 localhost:8080/login，这正是 Spring Security 的默认登录页，只要输入正确的用户名和密码便可跳转回原访问地址。

2.2 自定义表单登录页

1. 初步配置自定义表单登录页

虽然自动生成的表单登录页可以方便、快速地启动，但是大多数应用程序更希望提供自己的表单登录页，此时就需要覆写 configure 方法。

```
@EnableWebSecurity
public class WebSecurityConfig extends WebSecurityConfigurerAdapter {

    @Override
    protected void configure(HttpSecurity http) throws Exception {
        http.authorizeRequests()
                .anyRequest().authenticated()
                .and()
            .formLogin()
                .loginPage("/myLogin.html")
                // 使登录页不设限访问
                .permitAll()
                .and()
            .csrf().disable();
    }
}
```

2. 认识 HttpSecurity

HttpSecurity 实际上对应了 Spring Security 命名空间配置方式中 XML 文件内的标签，允许我们为特定的 HTTP 请求配置安全策略。

在 XML 文件中，声明大量配置早已司空见惯；但在 Java 配置中，按照传统的方式，我们需要这样来调用。

```
@EnableWebSecurity
public class WebSecurityConfig extends WebSecurityConfigurerAdapter {

    @Override
    protected void configure(HttpSecurity http) throws Exception {
```

```java
        ExpressionUrlAuthorizationConfigurer.ExpressionInterceptUrlRegistry
            urlRegistry = http.authorizeRequests();
        ExpressionUrlAuthorizationConfigurer.AuthorizedUrl authorizedUrl =
            (ExpressionUrlAuthorizationConfigurer.AuthorizedUrl)urlRegistry.
            anyRequest();
        authorizedUrl.authenticated();
        // more

        FormLoginConfigurer<HttpSecurity> formLoginConfigurer =
            http.formLogin();
        formLoginConfigurer.loginPage("/myLogin.html");
        formLoginConfigurer.permitAll();

        // more
    }
}
```

可以想象出这是多么烦琐且令人痛苦的一件事。HttpSecurity 首先被设计为链式调用，在执行每个方法后，都会返回一个预期的上下文，便于连续调用。我们不需要关心每个方法究竟返回了什么、如何进行下一个配置等细节。

HttpSecurity 提供了很多配置相关的方法，分别对应命名空间配置中的子标签<http>。例如，authorizeRequests()、formLogin()、httpBasic()和 csrf()分别对应<intercept-url>、<form-login>、<http-basic>和<csrf>标签。调用这些方法之后，除非使用 and()方法结束当前标签，上下文才会回到 HttpSecurity，否则链式调用的上下文将自动进入对应标签域。

authorizeRequests()方法实际上返回了一个 URL 拦截注册器，我们可以调用它提供的 anyanyRequest()、antMatchers()和 regexMatchers()等方法来匹配系统的 URL，并为其指定安全策略。

formLogin()方法和 httpBasic()方法都声明了需要 Spring Security 提供的表单认证方式，分别返回对应的配置器。其中，formLogin().loginPage("/myLogin.html")指定自定义的登录页/myLogin.html，同时，Spring Security 会用/myLogin.html 注册一个 POST 路由，用于接收登录请求。

csrf()方法是 Spring Security 提供的跨站请求伪造防护功能，当我们继承 WebSecurityConfigurerAdapter 时会默认开启 csrf()方法。关于 csrf()方法的更多内容会在后面的章节专门探讨，这里暂不讨论，以使测试进程更加顺利。

重新启动服务后再次访问 localhost:8080，页面会自动跳转到 localhost:8080/myLogin.html。由于/myLogin.html 无法定位到页面资源，所以会显示一个 404 页面，如图 2-4 所示。

图 2-4

3. 编写表单登录页

表单登录页的代码如下所示。

```html
<!DOCTYPE HTML>
<html>
    <head>
        <title>登录</title>
        <meta http-equiv="Content-Type" content="text/html; charset=utf-8" />
        <style>
            <!-- some style -->
        </style>
    </head>
    <body>
        <div class="login">
            <h2>Acced Form</h2>
            <div class="login-top">
                <h1>LOGIN FORM</h1>
                <form action="myLogin.html" method="post">
                    <input type="text" name="username" placeholder="username" />
                    <input type="password" name="password" placeholder="password" />
                    <div class="forgot">
                        <a href="#">forgot Password</a>
                        <input type="submit" value="Login" >
                    </div>
                </form>
            </div>
            <div class="login-bottom">
                <h3>New User  <a href="#">Register</a>  Here</h3>
            </div>
        </div>
```

```
</body>
</html>
```

myLogin.html 是开源模板，这里仅对其进行一些简单修改。因为篇幅有限，所以 CSS 样式部分已被截取。

在表单登录页中，仅有一个表单，用户名和密码分别为 username 和 password，并以 POST 的方式提交到/myLogin.html。

我们将其命名为 myLogin.html，放置在 resources/static/ 下。重启服务，再次访问 localhost:8080，即可看到自定义的表单登录页，如图 2-5 所示。

图 2-5

输入正确的用户名和密码后，单击"Login"按钮，即可成功跳转。

4．其他表单配置项

在自定义表单登录页之后，处理登录请求的 URL 也会相应改变。如何自定义 URL 呢？很简单，Spring Security 在表单定制里提供了相应的支持。

```
@EnableWebSecurity
public class WebSecurityConfig extends WebSecurityConfigurerAdapter {
    @Override
    protected void configure(HttpSecurity http) throws Exception {
        http.authorizeRequests()
```

```
            .anyRequest().authenticated()
            .and()
        .formLogin()
            .loginPage("/myLogin.html")
            // 指定处理登录请求的路径
            .loginProcessingUrl("/login")
            .permitAll()
            .and()
        .csrf().disable();
    }

}
```

此时，有些读者可能会有疑问，因为按照惯例，在发送登录请求并认证成功之后，页面会跳转回原访问页。在某些系统中的确是跳转回原访问页的，但在部分前后端完全分离、仅靠 JSON 完成所有交互的系统中，一般会在登录时返回一段 JSON 数据，告知前端登录成功与否，由前端决定如何处理后续逻辑，而非由服务器主动执行页面跳转。这在 Spring Security 中同样可以实现。

```
@EnableWebSecurity
public class WebSecurityConfig extends WebSecurityConfigurerAdapter {

    @Override
    protected void configure(HttpSecurity http) throws Exception {
        http.authorizeRequests()
            .anyRequest().authenticated()
            .and()
        .formLogin()
            .loginPage("/myLogin.html")
            // 指定处理登录请求的路径
            .loginProcessingUrl("/login")
            // 指定登录成功时的处理逻辑
            .successHandler(new AuthenticationSuccessHandler() {
                @Override
                public void onAuthenticationSuccess(HttpServletRequest
                    httpServletRequest, HttpServletResponse
                    httpServletResponse, Authentication authentication)
                    throws IOException, ServletException {
httpServletResponse.setContentType("application/json;charset=UTF-8");
                    PrintWriter out = httpServletResponse.getWriter();
                    out.write("{\"error_code\":\"0\", \"message\":\"欢迎登录系
```

```
                        统\"}");
                }
            })
            // 指定登录失败时的处理逻辑
            .failureHandler(new AuthenticationFailureHandler() {
                @Override
                public void onAuthenticationFailure(HttpServletRequest
                    httpServletRequest, HttpServletResponse
                    httpServletResponse, AuthenticationException e) throws
                    IOException, ServletException {
httpServletResponse.setContentType("application/json;charset=UTF-8");
                    httpServletResponse.setStatus(401);
                    PrintWriter out = httpServletResponse.getWriter();
                    // 输出失败的原因
                    out.write("{\"error_code\":\"401\", \"name\":\"" +
e.getClass() + "\", \"message\":\"" + e.getMessage() + "\"}");
                }
            })
            .permitAll()
            .and()
        .csrf().disable();
    }
}
```

表单登录配置模块提供了 successHandler() 和 failureHandler() 两个方法，分别处理登录成功和登录失败的逻辑。其中，successHandler() 方法带有一个 Authentication 参数，携带当前登录用户名及其角色等信息；而 failureHandler() 方法携带一个 AuthenticationException 异常参数。具体处理方式需按照系统的情况自定义。

在形式上，我们确实使用了 Spring Security 的表单认证功能，并且自定义了表单登录页。但实际上，这还远远不够。例如，在实际系统中，我们正常登录时使用的用户名和密码都来自数据库，这里却都写在配置上。更进一步，我们可以对每个登录用户都设定详细的权限，而并非一个通用角色。这些内容将在后面章节讲解。

第3章 认证与授权

在第2章中,我们沿用了 Spring Security 默认的安全机制:仅有一个用户,仅有一种角色。在实际开发中,这自然是无法满足需求的。本章将更加深入地对 Spring Security 进行配置,且初步使用授权机制。

3.1 默认数据库模型的认证与授权

3.1.1 资源准备

首先,在 controller 包下新建三个控制器,如图 3-1 所示。

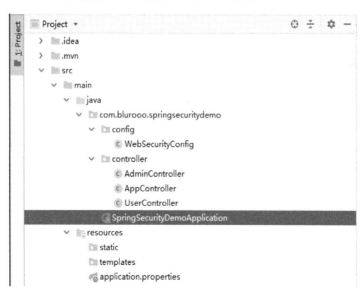

图 3-1

其次，分别建立一些测试路由。

```java
@RestController
@RequestMapping("/admin/api")
public class AdminController {

    @GetMapping("hello")
    public String hello() {
        return "hello, admin";
    }

}

@RestController
@RequestMapping("/app/api")
public class AppController {

    @GetMapping("hello")
    public String hello() {
        return "hello, app";
    }

}

@RestController
@RequestMapping("/user/api")
public class UserController {

    @GetMapping("hello")
    public String hello() {
        return "hello, user";
    }

}
```

假设在/admin/api下的内容是系统后台管理相关的 API，在/app/api 下的内容是面向客户端公开访问的 API，在/user/api下的内容是用户操作自身数据相关的 API；显然，/admin/api 必须拥有管理员权限才能进行操作，而/user/api 必须在用户登录后才能进行操作。

3.1.2 资源授权的配置

为了能正常访问前面的路由，我们需要进一步地配置 Spring Security。

```java
@EnableWebSecurity
public class WebSecurityConfig extends WebSecurityConfigurerAdapter {

    @Override
    protected void configure(HttpSecurity http) throws Exception {
        http.authorizeRequests()
                .antMatchers("/admin/api/**").hasRole("ADMIN")
                .antMatchers("/user/api/**").hasRole("USER")
                .antMatchers("/app/api/**").permitAll()
                .anyRequest().authenticated()
                .and()
            .formLogin().permitAll()
            .and()
        .csrf().disable();;
    }
}
```

　　antMatchers()是一个采用 ANT 模式的 URL 匹配器。ANT 模式使用?匹配任意单个字符，使用*匹配 0 或任意数量的字符，使用**匹配 0 或者更多的目录。antMatchers("/admin/api/**")相当于匹配了/admin/api/下的所有 API。此处我们指定当其必须为 ADMIN 角色时才能访问，/user/api/与之同理。/app/api/下的 API 会调用 permitAll()公开其权限。

　　授权相关的配置看起来并不复杂，但似乎缺少了什么？这里暂且忽略。

　　重启服务，尝试访问 localhost:8080/app/api/hello，页面打印"hello, app"，验证了/app/api/下的服务确实是权限公开的。接着访问 localhost:8080/user/api/hello，这次需要登录了。我们尝试输入前面在 application.properties 中定义的用户名和密码，登录之后，如果是 Spring Security 4.x 版本，则页面将会正常打印"hello, user"。然而，我们并没有角色为"USER"的用户，为什么可以成功访问路由呢？为了验证不是授权环节出现了问题，我们尝试访问 localhost:8080/admin/api/hello，如图 3-2 所示。

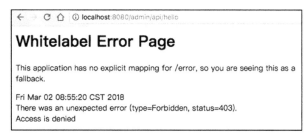

图 3-2

页面显示 403 错误,表示该用户授权失败(401 代表该用户认证失败)。也就是说,本次访问已经通过了认证环节,只是在授权的时候被驳回了。认证环节是没有问题的,其实是因为 Spring Security 4.x 默认的用户权限正是"ROLE_USER",如果对源码没有足够的了解,则可能会感到困惑,在 5.x 版本中就不存在这个问题了,用户已经不再有默认权限。

HTTP 状态码(HTTP Status Code)是由 RFC 2616 定义的一种用来表示一个 HTTP 请求响应状态的规范,由 3 位数字组成。通常用 2XX 表示本次操作成功,用 4XX 表示是客户端导致的失败,用 5XX 表示是服务器引起的错误。

3.1.3 基于内存的多用户支持

到目前为止,我们仍然只有一个可登录的用户,怎样引入多用户呢?非常简单,我们只需实现一个自定义的 UserDetailsService 即可。

```
@Bean
public UserDetailsService userDetailsService() {
   InMemoryUserDetailsManager manager = new InMemoryUserDetailsManager();
manager.createUser(User.withUsername("user").password("123").roles("USER").build());

manager.createUser(User.withUsername("admin").password("123").roles("USER",
"ADMIN").build());
   return manager;
}

// 由于 5.x 版本之后默认启用了委派密码编码器,
// 因而按照以往的方式设置内存密码将会读取异常,
// 所以需要暂时将密码编码器设置为 NoOpPasswordEncoder。
@Bean
public PasswordEncoder passwordEncoder() {
   return NoOpPasswordEncoder.getInstance();
}
```

为其添加一个@bean 注解,便可被 Spring Security 发现并使用。Spring Security 支持各种来源的用户数据,包括内存、数据库、LDAP 等。它们被抽象为一个 UserDetailsService 接口,任何实现了 UserDetailsService 接口的对象都可以作为认证数据源。在这种设计模式下,Spring Security 显得尤为灵活。

InMemoryUserDetailsManager 是 UserDetailsService 接口中的一个实现类，它将用户数据源寄存在内存里，在一些不需要引入数据库这种重数据源的系统中很有帮助。

这里仅仅调用 createUser() 生成两个用户，并赋予相应的角色。它会工作得很好，多次重启服务也不会出现问题。为什么要强调多次重启服务呢？稍后揭晓答案。

3.1.4　基于默认数据库模型的认证与授权

除了 InMemoryUserDetailsManager，Spring Security 还提供另一个 UserDetailsService 实现类：JdbcUserDetailsManager。

JdbcUserDetailsManager 帮助我们以 JDBC 的方式对接数据库和 Spring Security，它设定了一个默认的数据库模型，只要遵从这个模型，在简便性上，JdbcUserDetailsManager 甚至可以媲美 InMemoryUserDetailsManager。

1. 数据库准备

MySQL 的安装这里不赘述，首先在工程中引入 JDBC 和 MySQL 两个必要依赖。

```xml
<dependency>
        <groupId>org.springframework.boot</groupId>
        <artifactId>spring-boot-starter-jdbc</artifactId>
</dependency>
<dependency>
        <groupId>mysql</groupId>
        <artifactId>mysql-connector-java</artifactId>
        <scope>runtime</scope>
</dependency>
```

接着在 application.properties 中配置数据库连接参数。

```
spring.datasource.url=jdbc:mysql://localhost:3306/springDemo?useUnicode=true&characterEncoding=utf-8&useSSL=false&serverTimezone=GMT
spring.datasource.username=root
spring.datasource.password=123456
```

这里连接的数据库名为 springDemo（不配置 driverClassName 也不会出现问题，因为 Spring Boot 会自动根据 URL 去推断），用户名、密码分别为 root 和 123456，读者可根据实际情况，自行修改。

前面介绍过，JdbcUserDetailsManager 设定了一个默认的数据库模型，Spring Security 将该

模型定义在 /org/springframework/security/core/userdetails/jdbc/users.ddl 内。

```
create table users (
    username varchar_ignorecase(50) not null primary key,
    password varchar_ignorecase(500) not null,
    enabled boolean not null
);
create table authorities (
    username varchar_ignorecase(50) not null,
    authority varchar_ignorecase(50) not null,
    constraint fk_authorities_users foreign key(username) references users(username)
);
create unique index ix_auth_username on authorities (username,authority);
```

JdbcUserDetailsManager 需要两个表，其中 users 表用来存放用户名、密码和是否可用三个信息，authorities 表用来存放用户名及其权限的对应关系。

将其复制到 MySQL 命令窗口执行时，会报错，因为该语句是用 hsqldb 创建的，而 MySQL 不支持 varchar_ignorecase 这种类型。怎么办呢？很简单，将 varchar_ignorecase 改为 MySQL 支持的 varchar 即可。

```
mysql> create table users (
    ->     username varchar(50) not null primary key,
    ->     password varchar(500) not null,
    ->     enabled boolean not null
    -> );
Query OK, 0 rows affected (0.21 sec)

mysql> create table authorities (
    ->     username varchar(50) not null,
    ->     authority varchar(50) not null,
    ->     constraint fk_authorities_users foreign key(username) references users(userna
me)
    -> );
Query OK, 0 rows affected (0.16 sec)

mysql> create unique index ix_auth_username on authorities (username,authority);
Query OK, 0 rows affected (0.50 sec)
Records: 0  Duplicates: 0  Warnings: 0
```

2. 编码实现

下面构建一个 JdbcUserDetailsManager 实例，让 Spring Security 使用数据库来管理用户。

```
import org.apache.tomcat.jdbc.pool.DataSource;

@Autowired
private DataSource dataSource;
```

```java
@Bean
public UserDetailsService userDetailsService() {
    JdbcUserDetailsManager manager = new JdbcUserDetailsManager();
    manager.setDataSource(dataSource);
manager.createUser(User.withUsername("user").password("123").roles("USER").build());

manager.createUser(User.withUsername("admin").password("123").roles("USER","ADMIN").build());
    return manager;
}
```

JdbcUserDetailsManager 与 InMemoryUserDetailsManager 在用法上没有太大区别，只是多了设置 DataSource 的环节。Spring Security 通过 DataSource 执行设定好的命令。例如，此处的 createUser 函数实际上就是执行了下面的 SQL 语句。

```
insert into users (username, password, enabled) values (?,?,?)
```

查看 JdbcUserDetailsManager 的源代码可以看到更多定义好的 SQL 语句，诸如 deleteUserSql、updateUserSql 等，这些都是 JdbcUserDetailsManager 与数据库实际交互的形式。当然，JdbcUserDetailsManager 也允许我们在特殊情况下自定义这些 SQL 语句，如有必要，调用对应的 setXxxSql 方法即可。

现在重启服务，看看在数据库中 Spring Security 生成了哪些数据，如图 3-3 所示。

```
mysql> select * from users;
+----------+----------+---------+
| username | password | enabled |
+----------+----------+---------+
| admin    | 123      |       1 |
| user     | 123      |       1 |
+----------+----------+---------+

mysql> select * from authorities;
+----------+------------+
| username | authority  |
+----------+------------+
| user     | ROLE_USER  |
| admin    | ROLE_ADMIN |
| admin    | ROLE_USER  |
+----------+------------+
```

图 3-3

authorities 表的 authority 字段存放的是前面设定的角色，只是会被添上"ROLE_"前缀。下面尝试通过 SQL 命令创建一个测试账号。

```
insert into users values("test", "123", true);
insert into authorities values("test", "ROLE_USER");
```

清空缓存并使用测试账号访问系统，发现可以访问 user 路由，但不能访问 admin 路由，与预期的行为一致。

到目前为止，一切都工作得很好，但是只要我们重启服务，应用就会报错。这是因为 users 表在创建语句时，username 字段为主键，主键是唯一不重复的，但重启服务后会再次创建 admin 和 user，导致数据库报错（在内存数据源上不会出现这种问题，因为重启服务后会清空 username 字段中的内容）。所以如果需要在服务启动时便生成部分用户，那么建议先判断用户名是否存在。

```
@Bean
public UserDetailsService userDetailsService() {
    JdbcUserDetailsManager manager = new JdbcUserDetailsManager();
    manager.setDataSource(dataSource);
    if (!manager.userExists("user")) {

manager.createUser(User.withUsername("user").password("123").roles("USER").build());
    }
    if (!manager.userExists("admin")) {

manager.createUser(User.withUsername("admin").password("123").roles("USER", "ADMIN").build());
    }
    return manager;
}
```

在 2.2 节的自定义表单登录页中，WebSecurityConfigurer Adapter 定义了三个 configure。

```
protected void configure(AuthenticationManagerBuilder auth) throws Exception {
    this.disableLocalConfigureAuthenticationBldr = true;
}

public void configure(WebSecurity web) throws Exception {

}

protected void configure(HttpSecurity http) throws Exception {
        logger.debug("Using default configure(HttpSecurity). If subclassed this will potentially override subclass configure(HttpSecurity).");
```

```
        http
                .authorizeRequests()
                        .anyRequest().authenticated()
                        .and()
                .formLogin().and()
                .httpBasic();
}
```

我们只用到了一个参数，用来接收 HttpSecurity 对象的配置方法。另外两个参数也有各自的用途，其中，AuthenticationManagerBuilder 的 configure 同样允许我们配置认证用户。

```
@EnableWebSecurity
public class WebSecurityConfig extends WebSecurityConfigurerAdapter {

    @Override
    protected void configure(HttpSecurity http) throws Exception {
        http.authorizeRequests()
                .antMatchers("/admin/api/**").hasRole("ADMIN")
                .antMatchers("/user/api/**").hasRole("USER")
                .antMatchers("/app/api/**").permitAll()
                .anyRequest().authenticated()
                .and()
            .formLogin();
    }

    @Override
    protected void configure(AuthenticationManagerBuilder auth) throws Exception {
        auth.inMemoryAuthentication()
                .withUser("user").password("123").roles("user")
                .and()
                .withUser("admin").password("123").roles("admin");
    }

    @Bean
    public PasswordEncoder passwordEncoder() {
        return NoOpPasswordEncoder.getInstance();
    }

}
```

使用方法大同小异，这里不再赘述。

当使用 Spring Security 默认数据库模型应对各种用户系统时，难免灵活性欠佳。尤其是在对现有的系统做 Spring Security 嵌入时，原本的用户数据已经固定，为了适配 Spring Security 而在数据库层面进行修改显然得不偿失。强大而灵活的 Spring Security 对这方面有很好的支持。

3.2 自定义数据库模型的认证与授权

让 Spring Security 适应系统，而非让系统适应 Spring Security，是 Spring Security 框架开发者和使用者的共识。

下面我们将使用一个简单的自定义数据库模型接入 Spring Security，数据库依然是 MySQL，持久层框架则选用 MyBatis（倾向于使用 JPA 的读者也可以自行选型，它们在 Spring Security 部分的实践是一样的）。旁枝末节的知识会点到即止，我们重点介绍 Spring Security 相关的内容，所以期望读者自行阅读相关资料，也可以选择暂时略过。

3.2.1 实现 UserDetails

在 3.1 节，我们使用了 InMemoryUserDetailsManager 和 JdbcUserDetailsManager 两个 UserDetailsService 实现类。生效方式也很简单，只需加入 Spring 的 IoC 容器，就会被 Spring Security 自动发现并使用。自定义数据库结构实际上也仅需实现一个自定义的 UserDetailsService。

UserDetailsService 仅定义了一个 loadUserByUsername 方法，用于获取一个 UserDetails 对象。UserDetails 对象包含了一系列在验证时会用到的信息，包括用户名、密码、权限以及其他信息，Spring Security 会根据这些信息判定验证是否成功。

```
public interface UserDetails extends Serializable {
    Collection<? extends GrantedAuthority> getAuthorities();

    String getPassword();

    String getUsername();

    boolean isAccountNonExpired();

    boolean isAccountNonLocked();

    boolean isCredentialsNonExpired();
```

```
    boolean isEnabled();
}
```

也就是说，不管数据库结构如何变化，只要能构造一个 UserDetails 即可，下面就来实现这个过程。

1. 数据库准备

设计一个自定义的数据库结构。

```
CREATE TABLE 'users' (
  'id' bigint(20) NOT NULL AUTO_INCREMENT,
  'username' varchar(50) NOT NULL,
  'password' varchar(60),
  'enable' tinyint(4) NOT NULL DEFAULT '1' COMMENT '用户是否可用',
  'roles' text CHARACTER SET utf8 COMMENT '用户角色，多个角色之间用逗号隔开',
  PRIMARY KEY ('id'),
  KEY 'username' ('username')
);
```

我们把用户信息和角色放在同一张表中，不再是 Spring Security 默认的分开形式。roles 字段设定为 text 类型，多个角色之间用逗号隔开。建议在 username 字段上建立索引，以提高搜索速度，如图 3-4 所示。

```
mysql> desc users;
+----------+-------------+------+-----+---------+----------------+
| Field    | Type        | Null | Key | Default | Extra          |
+----------+-------------+------+-----+---------+----------------+
| id       | bigint(20)  | NO   | PRI | NULL    | auto_increment |
| username | varchar(50) | NO   | MUL | NULL    |                |
| password | varchar(50) | NO   |     | NULL    |                |
| enable   | tinyint(4)  | NO   |     | 1       |                |
| roles    | text        | YES  |     | NULL    |                |
+----------+-------------+------+-----+---------+----------------+
5 rows in set (0.03 sec)
```

图 3-4

接下来插入两条记录，显示如图 3-5 所示。

```
insert into users(username, password, roles) values("admin", "123", "ROLE_ADMIN,ROLE_USER");
insert into users(username, password, roles) values("user", "123", "ROLE_USER");
```

```
mysql> select * from users;
+----+----------+----------+--------+---------------------+
| id | username | password | enable | roles               |
+----+----------+----------+--------+---------------------+
|  1 | admin    | 123      |      1 | ROLE_ADMIN,ROLE_USER |
|  2 | user     | 123      |      1 | ROLE_USER           |
+----+----------+----------+--------+---------------------+
2 rows in set (0.00 sec)
```

图 3-5

2. 编码实现

当数据库结构和数据准备完毕时，即可编写对应的 User 实体。

```
public class User {

    private Long id;

    private String username;

    private String password;

    private String roles;

    private boolean enable;

    // setter and getter

}
```

让 User 实体继承 UserDetails。

```
public class User implements UserDetails {

    private Long id;

    private String username;

    private String password;

    private String roles;

    private boolean enable;

    // setter and getter
```

```java
    private List<GrantedAuthority> authorities;

    @Override
    public boolean isAccountNonExpired() {
        return true;
    }

    @Override
    public boolean isAccountNonLocked() {
        return true;
    }

    @Override
    public boolean isCredentialsNonExpired() {
        return true;
    }

    @Override
    public boolean isEnabled() {
        return this.enable;
    }

    public void setAuthorities(List<GrantedAuthority> authorities) {
        this.authorities = authorities;
    }

    @Override
    public Collection<? extends GrantedAuthority> getAuthorities() {
        return this.authorities;
    }
}
```

实现 UserDetails 定义的几个方法：

◎ isAccountNonExpired、isAccountNonLocked 和 isCredentialsNonExpired 暂且用不到，统一返回 true，否则 Spring Security 会认为账号异常。

◎ isEnabled 对应 enable 字段，将其代入即可。

◎ getAuthorities 方法本身对应的是 roles 字段，但由于结构不一致，所以此处新建一个，并在后续进行填充。

3.2.2 实现 UserDetailsService

1. 数据持久层准备

当准备好 UserDetails 之后,使用数据库持久层框架读取数据并填充对象。首先引入 MyBatis。

```xml
<dependency>
        <groupId>org.mybatis.spring.boot</groupId>
        <artifactId>mybatis-spring-boot-starter</artifactId>
        <version>1.3.1</version>
</dependency>
```

前面在配置文件中曾写过数据库相关的配置,这里沿用即可。

```
spring.datasource.url=jdbc:mysql://localhost:3306/springDemo?useUnicode=true
&characterEncoding=utf-8&useSSL=false&serverTimezone=GMT
spring.datasource.username=root
spring.datasource.password=123456
```

接下来在入口类中用@MapperScan 指定 MyBatis 要扫描的映射文件目录。

```java
@SpringBootApplication
@MapperScan("com.blurooo.springsecuritydemo.mapper")
public class SpringDemoApplication {

    public static void main(String[] args) {
        SpringApplication.run(SpringDemoApplication.class, args);
    }
}
```

当然,我们还需要在 com.blurooo.springsecuritydemor 下创建 mapper 目录,并编写对应的映射接口。

```java
@Component
public interface UserMapper {

  @Select("SELECT * FROM users WHERE username=#{username}")
  User findByUserName(@Param("username") String username);

}
```

与 MyBatis 相关的内容不再赘述,没有接触过且感兴趣的读者可以自行学习相关知识,这里仅提供一个通过用户名查找用户的方法。

2. 编码实现

当数据持久层准备完成后,我们开始编写 UserDetailsService。

```java
@Service
public class MyUserDetailsService implements UserDetailsService {

    @Autowired
    private UserMapper userMapper;

    @Override
    public UserDetails loadUserByUsername(String username) throws UsernameNotFoundException {
        // 从数据库尝试读取该用户
        User user = userMapper.findByUserName(username);
        // 用户不存在,抛出异常
        if (user == null) {
            throw new UsernameNotFoundException("用户不存在");
        }
        // 将数据库形式的 roles 解析为 UserDetails 的权限集
        // AuthorityUtils.commaSeparatedStringToAuthorityList 是 Spring Security
        // 提供的,该方法用于将逗号隔开的权限集字符串切割成可用权限对象列表
        // 当然也可以自己实现,如用分号来隔开等,参考 generateAuthorities
        user.setAuthorities(AuthorityUtils.commaSeparatedStringToAuthorityList(user.getRoles()));
        return user;
    }

    // 自行实现权限的转换
    private List<GrantedAuthority> generateAuthorities(String roles) {
        List<GrantedAuthority> authorities = new ArrayList<>();
        String roleArray = roles.split(";");
        if (roles != null && !"".equals(roles)) {
            for (String role : roleArray) {
                authorities.add(new SimpleGrantedAuthority(role));
            }
        }
        return authorities;
    }
}
```

其中,SimpleGrantedAuthority 是 GrantedAuthority 的一个实现类。Spring Security 的权限几乎是用 SimpleGrantedAuthority 生成的,只要注意每种角色对应一个 GrantedAuthority 即可。另

外,一定要在自己的 UserDetailsService 实现类上加入@Service 注解,以便被 Spring Security 自动发现。

至此,我们就实现了 Spring Security 的自定义数据库结构认证。有些读者可能会有疑问,为什么在数据库中的角色总是要添加"ROLE"前缀,在配置时却并没有"ROLE"前缀呢?

```
@Override
protected void configure(HttpSecurity http) throws Exception {
   http.cors()
          .and()
   .authorizeRequests()
          .antMatchers("/admin/api/**").hasRole("ADMIN")
          .antMatchers("/user/api/**").hasRole("USER")
          ...
}
```

查看源码即可找到答案。

```
private static String hasRole(String role) {
   Assert.notNull(role, "role cannot be null");
   if(role.startsWith("ROLE_")) {
       throw new IllegalArgumentException("role should not start with 'ROLE_' since it is automatically inserted. Got '" + role + "'");
   } else {
       return "hasRole('ROLE_" + role + "')";
   }
}
```

如果不希望匹配这个前缀,那么改为调用 hasAuthority 方法即可。

第 2 部分

第 4 章
实现图形验证码

在验证用户名和密码之前,引入辅助验证可有效防范暴力试错,图形验证码就是简单且行之有效的一种辅助验证方式。下面将使用过滤器和自定义认证两种方式实现图形验证码功能。

4.1 使用过滤器实现图形验证码

验证码(CAPTCHA)的全称是 Completely Automated Public Turing test to tell Computers and Humans Apart,翻译过来就是"全自动区分计算机和人类的图灵测试"。通俗地讲,验证码就是为了防止恶意用户暴力重试而设置的。不管是用户注册、用户登录,还是论坛发帖,如果不加以限制,一旦某些恶意用户利用计算机发起无限重试,就很容易使系统遭受破坏。

4.1.1 自定义过滤器

在 Spring Security 中,实现验证码校验的方式有很多种,最简单的方式就是自定义一个专门处理验证码逻辑的过滤器,将其添加到 Spring Security 过滤器链的合适位置。当匹配到登录请求时,立刻对验证码进行校验,成功则放行,失败则提前结束整个验证请求。

说到 Spring Security 的过滤器,我们先回顾一下前面使用过的配置。

```
@Override
protected void configure(HttpSecurity http) throws Exception {
  http
    .authorizeRequests()
        .antMatchers("/admin/api/**").hasAuthority("ROLE_ADMIN")
        .antMatchers("/user/api/**").hasRole("USER")
        .antMatchers("/app/api/**").permitAll()
        .anyRequest().authenticated()
```

```
      .and()
    .csrf().disable()
    .formLogin().permitAll()
      .and()
    .sessionManagement()
      .maximumSessions(1);
}
```

HttpSecurity 实际上就是在配置 Spring Security 的过滤器链，诸如 CSRF、CORS、表单登录等，每个配置器对应一个过滤器。我们可以通过 HttpSecurity 配置过滤器的行为，甚至可以像 CRSF 一样直接关闭过滤器。例如，SessionManagement。

```
public SessionManagementConfigurer<HttpSecurity> sessionManagement() throws Exception {
     return (SessionManagementConfigurer)this.getOrApply(new SessionManagementConfigurer());
}
```

Spring Security 通过 SessionManagementConfigurer 来配置 SessionManagement 的行为。与 SessionManagementConfigurer 类似的配置器还有 CorsConfigurer、RememberMeConfigurer 等，它们都实现了 SecurityConfigurer 的标准接口。

```
public interface SecurityConfigurer<O, B extends SecurityBuilder<O>> {

   // 各个配置器的初始化方法
   void init(B var1) throws Exception;

   // 各个配置器被统一调用的配置方法
   void configure(B var1) throws Exception;
}
```

SessionManagementConfigurer 是在 configure 方法中将最终的 SessionManagementFilter 插入过滤器链来实现会话管理的。

```
public void configure(H http) throws Exception {
       SecurityContextRepository securityContextRepository = http
                    .getSharedObject(SecurityContextRepository.class);
       SessionManagementFilter sessionManagementFilter = new SessionManagementFilter(
                    securityContextRepository,
getSessionAuthenticationStrategy(http));
       if (this.sessionAuthenticationErrorUrl != null) {
              sessionManagementFilter.setAuthenticationFailureHandler(
```

```
                                new SimpleUrlAuthenticationFailureHandler(
        this.sessionAuthenticationErrorUrl));
    }
    InvalidSessionStrategy strategy = getInvalidSessionStrategy();
    if (strategy != null) {

    sessionManagementFilter.setInvalidSessionStrategy(strategy);
    }
    AuthenticationFailureHandler failureHandler =
getSessionAuthenticationFailureHandler();
    if (failureHandler != null) {

    sessionManagementFilter.setAuthenticationFailureHandler(failureHandler);
    }
    AuthenticationTrustResolver trustResolver = http
        .getSharedObject(AuthenticationTrustResolver.class);
    if (trustResolver != null) {
            sessionManagementFilter.setTrustResolver(trustResolver);
    }
    sessionManagementFilter = postProcess(sessionManagementFilter);

    http.addFilter(sessionManagementFilter);
    if (isConcurrentSessionControlEnabled()) {
            ConcurrentSessionFilter concurrentSessionFilter =
createConccurencyFilter(http);

            concurrentSessionFilter =
postProcess(concurrentSessionFilter);
            http.addFilter(concurrentSessionFilter);
    }
}
```

除 Spring Security 提供的过滤器外，我们还可以添加自己的过滤器以实现更多的安全功能，这些都可以在 HttpSecurity 中实现。

```
// 将自定义过滤器添加在指定过滤器之后
public HttpSecurity addFilterAfter(Filter filter, Class<? extends Filter> afterFilter) {
    this.comparator.registerAfter(filter.getClass(), afterFilter);
    return this.addFilter(filter);
}
```

```java
// 将自定义过滤器添加在指定过滤器之前
public HttpSecurity addFilterBefore(Filter filter, Class<? extends Filter> beforeFilter) {
    this.comparator.registerBefore(filter.getClass(), beforeFilter);
    return this.addFilter(filter);
}

// 添加一个过滤器，但必须是 Spring Security 自身提供的过滤器实例或其继承过滤器，
// 详见 FilterComparator 类
public HttpSecurity addFilter(Filter filter) {
        Class<? extends Filter> filterClass = filter.getClass();
        if (!comparator.isRegistered(filterClass)) {
                throw new IllegalArgumentException(
                                    "The Filter class "
                                            + filterClass.getName()
                                            + " does not have a registered order and cannot be added without a specified order. Consider using addFilterBefore or addFilterAfter instead.");
        }
        this.filters.add(filter);
        return this;
}
// 添加一个自定义过滤器在指定过滤器位置
public HttpSecurity addFilterAt(Filter filter, Class<? extends Filter> atFilter) {
    this.comparator.registerAt(filter.getClass(), atFilter);
    return this.addFilter(filter);
}
```

4.1.2 图形验证码过滤器

毋庸置疑，要想实现图形验证码校验功能，首先应当有一个用于获取图形验证码的 API。绘制图形验证码的方法有很多，使用开源的验证码组件即可，例如 kaptcha（请勿用在生产）。

```xml
<dependency>
        <groupId>com.github.penggle</groupId>
        <artifactId>kaptcha</artifactId>
        <version>2.3.2</version>
</dependency>
```

kaptcha 是一个很小的工具，使用方法极其简单，不再赘述。

首先配置一个 kaptcha 实例。

```
@Bean
public Producer captcha() {
    // 配置图形验证码的基本参数
    Properties properties = new Properties();
    // 图片宽度
    properties.setProperty("kaptcha.image.width", "150");
    // 图片长度
    properties.setProperty("kaptcha.image.height", "50");
    // 字符集
    properties.setProperty("kaptcha.textproducer.char.string", "0123456789");
    // 字符长度
    properties.setProperty("kaptcha.textproducer.char.length", "4");
    Config config = new Config(properties);
    // 使用默认的图形验证码实现，当然也可以自定义实现
    DefaultKaptcha defaultKaptcha = new DefaultKaptcha();
    defaultKaptcha.setConfig(config);
    return defaultKaptcha;
}
```

接着创建一个 CaptchaController，用于获取图形验证码。

```
@Controller
public class CaptchaController {

    @Autowired
    private Producer captchaProducer;

    @GetMapping("/captcha.jpg")
    public void getCaptcha(HttpServletRequest request, HttpServletResponse response) throws IOException {
        // 设置内容类型
        response.setContentType("image/jpeg");
        // 创建验证码文本
        String capText = captchaProducer.createText();
        // 将验证码文本设置到 session
        request.getSession().setAttribute("captcha", capText);
        // 创建验证码图片
        BufferedImage bi = captchaProducer.createImage(capText);
        // 获取响应输出流
        ServletOutputStream out = response.getOutputStream();
        // 将图片验证码数据写到响应输出流
```

```
        ImageIO.write(bi, "jpg", out);
        // 推送并关闭响应输出流
        try {
            out.flush();
        } finally {
            out.close();
        }
    }
}
```

当用户访问/captcha.jpg 时，即可得到一张携带验证码的图片，验证码文本则被存放到 session 中，用于后续的校验。

有了图形验证码的 API 之后，就可以自定义验证码校验过滤器了。虽然 Spring Security 的过滤器链对过滤器没有特殊要求，只要继承了 Filter 即可，但是在 Spring 体系中，推荐使用 OncePerRequestFilter 来实现，它可以确保一次请求只会通过一次该过滤器（Filter 实际上并不能保证这一点）。

```
// 自定义一个验证码校验失败的异常
public class VerificationCodeException extends AuthenticationException {

    public VerificationCodeException () {
        super("图形验证码校验失败");
    }

}
// 专门用于校验验证码的过滤器
public class VerificationCodeFilter extends OncePerRequestFilter {

    private AuthenticationFailureHandler authenticationFailureHandler = new MyAuthenticationFailureHandler();


    @Override
    protected void doFilterInternal(HttpServletRequest httpServletRequest,
HttpServletResponse httpServletResponse, FilterChain filterChain) throws
ServletException, IOException {
        // 非登录请求不校验验证码
        if (!"/auth/form".equals(httpServletRequest.getRequestURI())) {
            filterChain.doFilter(httpServletRequest, httpServletResponse);
        } else {
```

```
            try {
                verificationCode(httpServletRequest);
                filterChain.doFilter(httpServletRequest, httpServletResponse);
            } catch (VerificationCodeException e) {
authenticationFailureHandler.onAuthenticationFailure(httpServletRequest,
httpServletResponse, e);
            }
        }
    }

    public void verificationCode (HttpServletRequest httpServletRequest) throws
VerificationCodeException {
        String requestCode = httpServletRequest.getParameter("captcha");
        HttpSession session = httpServletRequest.getSession();
        String savedCode = (String) session.getAttribute("captcha");
        if (!StringUtils.isEmpty(savedCode)) {
            // 随手清除验证码,无论是失败,还是成功。客户端应在登录失败时刷新验证码
            session.removeAttribute("captcha");
        }
        // 校验不通过,抛出异常
        if (StringUtils.isEmpty(requestCode) || StringUtils.isEmpty(savedCode)
|| !requestCode.equals(savedCode)) {
            throw new VerificationCodeException();
        }
    }
}
```

总体来说,验证码仅仅核对 session 中保存的验证码与用户提交的验证码是否一致,逻辑并不复杂,只需将该过滤器添加到 Spring Security 的过滤器链中即可生效。

```
@Override
protected void configure(HttpSecurity http) throws Exception {
    http
    .authorizeRequests()
            .antMatchers("/admin/api/**").hasAuthority("ROLE_ADMIN")
            .antMatchers("/user/api/**").hasRole("USER")
            // 开放 captcha.jpg 的访问权限
            .antMatchers("/app/api/**", "/captcha.jpg").permitAll()
            .anyRequest().authenticated()
            .and()
            .csrf().disable()
        .formLogin()
            .loginPage("/myLogin.html")
```

```
        .loginProcessingUrl("/auth/form").permitAll()
        .failureHandler(new MyAuthenticationFailureHandler());
// 将过滤器添加在 UsernamePasswordAuthenticationFilter 之前
http.addFilterBefore(new VerificationCodeFilter(),
UsernamePasswordAuthenticationFilter.class);
}
```

为了体验效果，下面修改自定义表单登录页。

```
<!DOCTYPE HTML>
<html>
    <head>
        <title>登录</title>
        <meta http-equiv="Content-Type" content="text/html; charset=utf-8" />
    </head>
    <body>
        <div class="login">
            <h2>Acced Form</h2>
            <div class="login-top">
                <h1>LOGIN FORM</h1>
                <form action="/auth/form" method="post">
                    <input type="text" name="username" placeholder="username" />
                    <input type="password" name="password" placeholder="password" />
                    <div style="display: flex;">
                        <!-- 新增图形验证码的输入框 -->
                        <input type="text" name="captcha" placeholder="captcha" />
                        <!-- 图片指向图形验证码 API -->
                        <img src="/captcha.jpg" alt="captcha" height="50px" width="150px" style="margin-left: 20px;">
                    </div>
                    <div class="forgot">
                        <a href="#">forgot Password</a>
                        <input type="submit" value="Login" >
                    </div>
                </form>
            </div>
            <div class="login-bottom">
                <h3>New User  <a href="#">Register</a>  Here</h3>
            </div>
        </div>
        <style>
            <!-- css 样式表省略 -->
        </style>
```

```
    </body>
</html>
```

重启服务，访问任意一个受保护的资源即可看到带验证码的登录页面，如图 4-1 所示。

当正确输入验证码时，可以正常访问对应资源。只要输入的验证码有误，就会返回一个 401 页面，如图 4-2 所示。

图 4-1   图 4-2

至此，验证码登录功能就粗略实现了，而定制验证码的尺寸、验证码的失效属性、配置验证码的字符取值范围等，可以在此基础上集成。

4.2　使用自定义认证实现图形验证码

前面使用过滤器的方式实现了带图形验证码的验证功能，属于 Servlet 层面，简单、易理解。其实，Spring Security 还提供了一种更优雅的实现图形验证码的方式，即自定义认证。

4.2.1　认识 AuthenticationProvider

在学习 Spring Security 的自定义认证之前，有必要了解 Spring Security 是如何灵活集成多种认证技术的。

我们所面对的系统中的用户，在 Spring Security 中被称为主体（principal）。主体包含了所有能够经过验证而获得系统访问权限的用户、设备或其他系统。主体的概念实际上来自 Java

Security，Spring Security 通过一层包装将其定义为一个 Authentication。

```
public interface Authentication extends Principal, Serializable {

    // 获取主体权限列表
    Collection<? extends GrantedAuthority> getAuthorities();

    // 获取主体凭据，通常为用户密码
    Object getCredentials();

    // 获取主体携带的详细信息
    Object getDetails();

    // 获取主体，通常为一个用户名
    Object getPrincipal();

    // 主体是否验证成功
    boolean isAuthenticated();

    void setAuthenticated(boolean var1) throws IllegalArgumentException;
}
```

Authentication 中包含主体权限列表、主体凭据、主体详细信息，以及主体是否验证成功等信息。由于大部分场景下身份验证都是基于用户名和密码进行的，所以 Spring Security 提供了一个 UsernamePasswordAuthenticationToken 用于代指这一类证明（例如，用 SSH KEY 也可以登录，但它不属于用户名和密码登录这个范畴，如有必要，也可以自定义提供）。在前面使用的表单登录中，每一个登录用户都被包装为一个 UsernamePasswordAuthenticationToken，从而在 Spring Security 的各个 AuthenticationProvider 中流动。

AuthenticationProvider 被 Spring Security 定义为一个验证过程。

```
public interface AuthenticationProvider {

    // 验证过程，成功返回一个验证完成的 Authentication
    Authentication authenticate(Authentication var1) throws AuthenticationException;

    // 是否支持验证当前的 Authentication 类型
    boolean supports(Class<?> var1);
}
```

一次完整的认证可以包含多个 AuthenticationProvider，一般由 ProviderManager 管理。

```java
public class ProviderManager implements AuthenticationManager,
MessageSourceAware, InitializingBean {

    private List<AuthenticationProvider> providers;

    // 验证
    public Authentication authenticate(Authentication authentication) throws
AuthenticationException {
        Class<? extends Authentication> toTest = authentication.getClass();
        AuthenticationException lastException = null;
        Authentication result = null;
        boolean debug = logger.isDebugEnabled();
        Iterator var6 = this.getProviders().iterator();

        // 迭代验证每个AuthenticationProvider，直到有一个验证通过，即可跳出
        while(var6.hasNext()) {
            AuthenticationProvider provider =
(AuthenticationProvider)var6.next();
            if(provider.supports(toTest)) {
                if(debug) {
                    logger.debug("Authentication attempt using " +
provider.getClass().getName());
                }

                try {
                    result = provider.authenticate(authentication);
                    if(result != null) {
                        this.copyDetails(authentication, result);
                        break;
                    }
                } catch (AccountStatusException var11) {
                    this.prepareException(var11, authentication);
                    throw var11;
                } catch (InternalAuthenticationServiceException var12) {
                    this.prepareException(var12, authentication);
                    throw var12;
                } catch (AuthenticationException var13) {
                    lastException = var13;
                }
            }
        }

        if(result == null && this.parent != null) {
```

```
            try {
                result = this.parent.authenticate(authentication);
            } catch (ProviderNotFoundException var9) {
                ;
            } catch (AuthenticationException var10) {
                lastException = var10;
            }
        }

        if(result != null) {
            if(this.eraseCredentialsAfterAuthentication && result instanceof
CredentialsContainer) {
                ((CredentialsContainer)result).eraseCredentials();
            }

            this.eventPublisher.publishAuthenticationSuccess(result);
            return result;
        } else {
            if(lastException == null) {
                lastException = new
ProviderNotFoundException(this.messages.getMessage("ProviderManager.provider
NotFound", new Object[]{toTest.getName()}, "No AuthenticationProvider found for
{0}"));
            }

            this.prepareException((AuthenticationException)lastException,
authentication);
            throw lastException;
        }
    }

}
```

4.2.2　自定义 AuthenticationProvider

Spring Security 提供了多种常见的认证技术，包括但不限于以下几种：

◎　HTTP 层面的认证技术，包括 HTTP 基本认证和 HTTP 摘要认证两种。

◎　基于 LDAP 的认证技术（Lightweight Directory Access Protocol，轻量目录访问协议）。

◎　聚焦于证明用户身份的 OpenID 认证技术。

◎　聚焦于授权的 OAuth 认证技术。

◎ 系统内维护的用户名和密码认证技术。

其中，使用最为广泛的是由系统维护的用户名和密码认证技术，通常会涉及数据库访问。为了更好地按需定制，Spring Security 并没有直接糅合整个认证过程，而是提供了一个抽象的 AuthenticationProvider。

```
public abstract class AbstractUserDetailsAuthenticationProvider implements
AuthenticationProvider, InitializingBean, MessageSourceAware {

    // 附加认证过程
    protected abstract void additionalAuthenticationChecks(UserDetails var1,
UsernamePasswordAuthenticationToken var2) throws AuthenticationException;

    // 检索用户
    protected abstract UserDetails retrieveUser(String var1,
UsernamePasswordAuthenticationToken var2) throws AuthenticationException;


    // 认证过程
    public Authentication authenticate(Authentication authentication) throws
AuthenticationException {
        Assert.isInstanceOf(UsernamePasswordAuthenticationToken.class,
authentication,
this.messages.getMessage("AbstractUserDetailsAuthenticationProvider.onlySupp
orts", "Only UsernamePasswordAuthenticationToken is supported"));
        String username = authentication.getPrincipal() ==
null?"NONE_PROVIDED":authentication.getName();
        boolean cacheWasUsed = true;
        UserDetails user = this.userCache.getUserFromCache(username);
        if(user == null) {
            cacheWasUsed = false;

            try {
                // 先检索用户
                user = this.retrieveUser(username,
(UsernamePasswordAuthenticationToken)authentication);
            } catch (UsernameNotFoundException var6) {
                this.logger.debug("User '" + username + "' not found");
                if(this.hideUserNotFoundExceptions) {
                    throw new
BadCredentialsException(this.messages.getMessage("AbstractUserDetailsAuthent
icationProvider.badCredentials", "Bad credentials"));
                }
```

```
            throw var6;
        }

        Assert.notNull(user, "retrieveUser returned null - a violation of the
interface contract");
    }

    try {
        // 检查用户账号是否可用
        this.preAuthenticationChecks.check(user);
        // 附加认证
        this.additionalAuthenticationChecks(user,
(UsernamePasswordAuthenticationToken)authentication);
    } catch (AuthenticationException var7) {
        if(!cacheWasUsed) {
            throw var7;
        }

        cacheWasUsed = false;
        user = this.retrieveUser(username,
(UsernamePasswordAuthenticationToken)authentication);
        this.preAuthenticationChecks.check(user);
        this.additionalAuthenticationChecks(user,
(UsernamePasswordAuthenticationToken)authentication);
    }
    // 检查用户密码是否过期
    this.postAuthenticationChecks.check(user);
    if(!cacheWasUsed) {
        this.userCache.putUserInCache(user);
    }

    Object principalToReturn = user;
    if(this.forcePrincipalAsString) {
        principalToReturn = user.getUsername();
    }

    // 返回一个认证通过的 Authentication
    return this.createSuccessAuthentication(principalToReturn,
authentication, user);
}

// 此认证过程支持 UsernamePasswordAuthenticationToken 及衍生对象
```

```java
    public boolean supports(Class<?> authentication) {
        return UsernamePasswordAuthenticationToken.class.isAssignableFrom(authentication);
    }

    ...
}
```

在 AbstractUserDetailsAuthenticationProvider 中实现了基本的认证流程，通过继承 AbstractUserDetailsAuthenticationProvider，并实现 retrieveUser 和 additionalAuthenticationChecks 两个抽象方法即可自定义核心认证过程，灵活性非常高。

```java
@Component
public class MyAuthenticationProvider extends AbstractUserDetailsAuthenticationProvider {

    @Autowired
    private UserDetailsService userDetailsService;

    @Override
    protected void additionalAuthenticationChecks(UserDetails userDetails, UsernamePasswordAuthenticationToken usernamePasswordAuthenticationToken) throws AuthenticationException {
        // 编写更多校验逻辑

        // 校验密码
        if(usernamePasswordAuthenticationToken.getCredentials() == null) {
            throw new BadCredentialsException(this.messages.getMessage("AbstractUserDetailsAuthenticationProvider.badCredentials", "密码不能为空"));
        } else {
            String presentedPassword = usernamePasswordAuthenticationToken.getCredentials().toString();
            if(!presentedPassword.equals(userDetails.getPassword())) {
                throw new BadCredentialsException(this.messages.getMessage("AbstractUserDetailsAuthenticationProvider.badCredentials", "密码错误"));
            }
        }
    }

    @Override
    protected UserDetails retrieveUser(String s,
```

```
UsernamePasswordAuthenticationToken usernamePasswordAuthenticationToken)
throws AuthenticationException {
    return userDetailsService.loadUserByUsername(s);
    }
}
```

Spring Security 同样提供一个继承自 AbstractUserDetailsAuthenticationProvider 的 AuthenticationProvider。

```
public class DaoAuthenticationProvider extends
AbstractUserDetailsAuthenticationProvider {
    private static final String USER_NOT_FOUND_PASSWORD = "userNotFoundPassword";
    private PasswordEncoder passwordEncoder;
    private String userNotFoundEncodedPassword;
    private SaltSource saltSource;
    private UserDetailsService userDetailsService;

    public DaoAuthenticationProvider() {
        this.setPasswordEncoder((PasswordEncoder)(new
PlaintextPasswordEncoder()));
    }

    protected void additionalAuthenticationChecks(UserDetails userDetails,
UsernamePasswordAuthenticationToken authentication) throws
AuthenticationException {
        Object salt = null;
        if(this.saltSource != null) {
            salt = this.saltSource.getSalt(userDetails);
        }

        if(authentication.getCredentials() == null) {
            this.logger.debug("Authentication failed: no credentials provided");
            throw new
BadCredentialsException(this.messages.getMessage("AbstractUserDetailsAuthent
icationProvider.badCredentials", "Bad credentials"));
        } else {
            String presentedPassword =
authentication.getCredentials().toString();
            if(!this.passwordEncoder.isPasswordValid(userDetails.getPassword(),
presentedPassword, salt)) {
                this.logger.debug("Authentication failed: password does not match
stored value");
                throw new
BadCredentialsException(this.messages.getMessage("AbstractUserDetailsAuthent
```

```
icationProvider.badCredentials", "Bad credentials"));
            }
        }
    }

    protected final UserDetails retrieveUser(String username,
UsernamePasswordAuthenticationToken authentication) throws
AuthenticationException {
        UserDetails loadedUser;
        try {
            loadedUser =
this.getUserDetailsService().loadUserByUsername(username);
        } catch (UsernameNotFoundException var6) {
            if(authentication.getCredentials() != null) {
                String presentedPassword =
authentication.getCredentials().toString();

this.passwordEncoder.isPasswordValid(this.userNotFoundEncodedPassword,
presentedPassword, (Object)null);
            }

            throw var6;
        } catch (Exception var7) {
            throw new InternalAuthenticationServiceException(var7.getMessage(),
var7);
        }

        if(loadedUser == null) {
            throw new
InternalAuthenticationServiceException("UserDetailsService returned null,
which is an interface contract violation");
        } else {
            return loadedUser;
        }
    }
    ...
}
```

　　DaoAuthenticationProvider 的用户信息来源于 UserDetailsService，并且整合了密码编码的实现，在前面章节中学习的表单认证就是由 DaoAuthenticationProvider 提供的。

4.2.3 实现图形验证码的 AuthenticationProvider

前面我们已经基本了解了 Spring Security 的认证流程，现在重新回到自定义认证实现图形验证码登录这个具体案例中。由于只是在常规的认证之上增加了图形验证码的校验，其他流程并没有变化，所以只需继承 DaoAuthenticationProvider 并稍作增添即可。

```java
@Component
public class MyAuthenticationProvider extends DaoAuthenticationProvider {

    // 把构造方法注入 UserDetailService 和 PasswordEncoder
    public MyAuthenticationProvider(UserDetailsService userDetailsService,
 PasswordEncoder passwordEncoder) {
        this.setUserDetailsService(userDetailsService);
        this.setPasswordEncoder(passwordEncoder);
    }

    @Override
    protected void additionalAuthenticationChecks(UserDetails userDetails,
 UsernamePasswordAuthenticationToken usernamePasswordAuthenticationToken)
 throws AuthenticationException {
        // 实现图形验证码的校验逻辑

        // 调用父类方法完成密码验证
        super.additionalAuthenticationChecks(userDetails,
 usernamePasswordAuthenticationToken);
    }
}
```

在验证流程中添加新的逻辑后似乎有些问题。在 additionalAuthenticationChecks 中，我们可以得到的参数是来自 UserDetailsService 的 UserDetails，以及根据用户提交的账号信息封装而来的 UsernamePasswordAuthenticationToken，而图形验证码的校验必须要有 HttpServletRequest 对象，因为用户提交的验证码和 session 存储的验证码都需要从用户的请求中获取，这是否意味着这种实现方式不可行呢？并非如此，Authentication 实际上还可以携带账号信息之外的数据。

```java
public interface Authentication extends Principal, Serializable {

    // 允许携带任意对象
    Object getDetails();
```

}

如果这个数据可以利用,那么难题自然就迎刃而解了。前面提到过,一次完整的认证可以包含多个 AuthenticationProvider,这些 AuthenticationProvider 都是由 ProviderManager 管理的,而 ProviderManager 是由 UsernamePasswordAuthenticationFilter 调用的。也就是说,所有的 AuthenticationProvider 包含的 Authentication 都来源于 UsernamePasswordAuthenticationFilter。

```java
public class UsernamePasswordAuthenticationFilter extends
AbstractAuthenticationProcessingFilter {

    ...

    public Authentication attemptAuthentication(HttpServletRequest request,
HttpServletResponse response) throws AuthenticationException {
        if(this.postOnly && !request.getMethod().equals("POST")) {
            throw new AuthenticationServiceException("Authentication method not
supported: " + request.getMethod());
        } else {
            String username = this.obtainUsername(request);
            String password = this.obtainPassword(request);
            if(username == null) {
                username = "";
            }

            if(password == null) {
                password = "";
            }

            username = username.trim();
            // 生成一个基本的 Authentication
            UsernamePasswordAuthenticationToken authRequest = new
UsernamePasswordAuthenticationToken(username, password);
            // 为该 Authentication 设置详细信息
            this.setDetails(request, authRequest);
            // 调用 ProviderManager,将该 Authentication 传入认证流程
            return this.getAuthenticationManager().authenticate(authRequest);
        }
    }

    ...

    protected void setDetails(HttpServletRequest request,
```

```
UsernamePasswordAuthenticationToken authRequest) {
    // 通过AuthenticationDetailsSource构建详细信息,携带一个HttpServletRequest
    // 对象
authRequest.setDetails(this.authenticationDetailsSource.buildDetails(request
));
    }
}
```

AbstractAuthenticationProcessingFilter 本身并没有设置用户详细信息的流程,而且是通过标准接口 AuthenticationDetailsSource 构建的,这意味着它是一个允许定制的特性。

```
public interface AuthenticationDetailsSource<C, T> {
   T buildDetails(C var1);
}
```

在 UsernamePasswordAuthenticationFilter 中使用的 AuthenticationDetailsSource 是一个标准的 Web 认证源,携带的是用户的 sessionId 和 IP 地址。

```
public class WebAuthenticationDetailsSource implements
AuthenticationDetailsSource<HttpServletRequest, WebAuthenticationDetails> {
   public WebAuthenticationDetailsSource() {
   }

   public WebAuthenticationDetails buildDetails(HttpServletRequest context) {
      return new WebAuthenticationDetails(context);
   }
}

public class WebAuthenticationDetails implements Serializable {
   private static final long serialVersionUID = 420L;
   private final String remoteAddress;
   private final String sessionId;

   public WebAuthenticationDetails(HttpServletRequest request) {
      this.remoteAddress = request.getRemoteAddr();
      HttpSession session = request.getSession(false);
      this.sessionId = session != null?session.getId():null;
   }

   ...

}
```

有了 HttpServletRequest 之后，一切都将变得非常顺畅。基于图形验证码的场景，我们可以继承 WebAuthenticationDetails，并扩展需要的信息。

```java
public class MyWebAuthenticationDetails extends WebAuthenticationDetails {

    private String imageCode;

    private String savedImageCode;

    public String getImageCode() {
        return imageCode;
    }

    public String getSavedImageCode() {
        return savedImageCode;
    }

    // 补充用户提交的验证码和 session 保存的验证码
    public MyWebAuthenticationDetails(HttpServletRequest request) {
        super(request);
        this.imageCode = request.getParameter("captcha");
        HttpSession session = request.getSession();
        this.savedImageCode = (String) session.getAttribute("captcha");
        if (!StringUtils.isEmpty(this.savedImageCode)) {
            // 随手清除验证码，不管是失败还是成功，所以客户端应在登录失败时刷新验证码
            session.removeAttribute("captcha");
        }
    }
}
```

将它提供给一个自定义的 AuthenticationDetailsSource。

```java
@Component
public class MyWebAuthenticationDetailsSource implements
AuthenticationDetailsSource<HttpServletRequest, WebAuthenticationDetails> {
    @Override
    public WebAuthenticationDetails buildDetails(HttpServletRequest request) {
        return new MyWebAuthenticationDetails(request);
    }
}
```

接下来实现我们自定义的 AuthenticationProvider。

```
@Component
public class MyAuthenticationProvider extends DaoAuthenticationProvider {

    // 构造方法注入 UserDetailService 和 PasswordEncoder
    public MyAuthenticationProvider(UserDetailsService userDetailsService,
PasswordEncoder passwordEncoder) {
        this.setUserDetailsService(userDetailsService);
        this.setPasswordEncoder(passwordEncoder);
    }

    @Override
    protected void additionalAuthenticationChecks(UserDetails userDetails,
UsernamePasswordAuthenticationToken usernamePasswordAuthenticationToken)
throws AuthenticationException {
        MyWebAuthenticationDetails details = (MyWebAuthenticationDetails)
usernamePasswordAuthenticationToken.getDetails();
        String imageCode = details.getImageCode();
        String savedImageCode = details.getSavedImageCode();
        // 检验图形验证码
        if (StringUtils.isEmpty(imageCode) ||
StringUtils.isEmpty(savedImageCode) || !imageCode.equals(savedImageCode)) {
            throw new VerificationCodeException();
        }
        super.additionalAuthenticationChecks(userDetails,
usernamePasswordAuthenticationToken);
    }

}
```

想要应用自定义的 AuthenticationProvider 和 AuthenticationDetailsSource，还需在 WebSecurityConfig 中完成剩余的配置。

```
@EnableWebSecurity
public class WebSecurityConfig extends WebSecurityConfigurerAdapter {

    @Autowired
    private AuthenticationDetailsSource<HttpServletRequest,
WebAuthenticationDetails> myWebAuthenticationDetailsSource;

    @Autowired
    private AuthenticationProvider authenticationProvider;
```

```java
    @Override
    protected void configure(AuthenticationManagerBuilder auth) throws 
Exception {
        // 应用 AuthenticationProvider
        auth.authenticationProvider(authenticationProvider);
    }

    @Override
    protected void configure(HttpSecurity http) throws Exception {
        http
            .authorizeRequests()
                .antMatchers("/admin/api/**").hasAuthority("ROLE_ADMIN")
                .antMatchers("/user/api/**").hasRole("USER")
                .antMatchers("/app/api/**", "/captcha.jpg").permitAll()
                .anyRequest().authenticated()
                .and()
            .csrf().disable()
            .formLogin()
                // 应用 AuthenticationDetailsSource
                .authenticationDetailsSource(myWebAuthenticationDetailsSource)
                .loginPage("/myLogin.html")
                .loginProcessingUrl("/auth/form").permitAll()
                .failureHandler(new MyAuthenticationFailureHandler())
                .and()
            .csrf().disable();
    }

    @Bean
    public PasswordEncoder passwordEncoder() {
        return NoOpPasswordEncoder.getInstance();
    }

}
```

第 5 章
自动登录和注销登录

关于网站的安全设计，通常是有一些矛盾点的。我们在作为某些系统开发者的同时，也在充当着另外一些系统的用户，一些感同身受的东西可以带来很多思考。

5.1 为什么需要自动登录

当我们在某个网站上注册账号时，网站会对我们设置的登录密码提出要求。例如，有的网站要求使用固定位数的纯数字密码，有的网站则强制要求用户使用英文+数字组合成的密码，甚至要求加一些特殊符号来组成密码。总体而言，设定一个密码并不困难，真正的困难总是在下次登录时才会遇到。要么想不出网站要求的密码格式是什么，要么还原不了设置密码时的思维状态。总之，在几次尝试登录失败之后，大部分人会选择找回密码，从而再次陷入如何设置密码的循环里。

为了尽可能减少用户重新登录的频率，在系统开发之初就需要考虑加入可以提升用户登录体验的功能。自动登录便是这样一个会给用户带来便利，同时也会给用户带来风险的体验性功能。

自动登录是将用户的登录信息保存在用户浏览器的 cookie 中，当用户下次访问时，自动实现校验并建立登录态的一种机制。

Spring Security 提供了两种非常好的令牌：

◎ 用散列算法加密用户必要的登录信息并生成令牌。
◎ 数据库等持久性数据存储机制用的持久化令牌。

散列算法在 Spring Security 中是通过加密几个关键信息实现的。

```
hashInfo = md5Hex(username + ":" + expirationTime + ":" password + ":" + key)
rememberCookie = base64(username + ":" + expirationTime + ":" + hashInfo)
```

其中，expirationTime 指本次自动登录的有效期，key 为指定的一个散列盐值，用于防止令牌被修改。通过这种方式生成 cookie 后，在下次登录时，Spring Security 首先用 Base64 简单解码得到用户名、过期时间和加密散列值；然后使用用户名得到密码；接着重新以该散列算法正向计算，并将计算结果与旧的加密散列值进行对比，从而确认该令牌是否有效。

5.2 实现自动登录

1. 散列加密方案

在 Spring Security 中加入自动登录的功能非常简单。

```
@Autowired
private MyUserDetailsService userDetailsService;

@Override
protected void configure(HttpSecurity http) throws Exception {
    http
    .authorizeRequests()
         .antMatchers("/admin/api/**").hasRole("ADMIN")
         .antMatchers("/user/api/**").hasRole("USER")
         .antMatchers("/app/api/**").permitAll()
         .anyRequest().authenticated()
         .and()
         .csrf().disable()
     .formLogin().permitAll()
         .and()
     // 增加自动登录功能，默认为简单散列加密
         .rememberMe().userDetailsService(userDetailsService);
}
```

前提是已经实现了一个 UserDetailsService。重启服务后访问受限 API，这次在表单登录页中多了一个可选框，如图 5-1 所示。

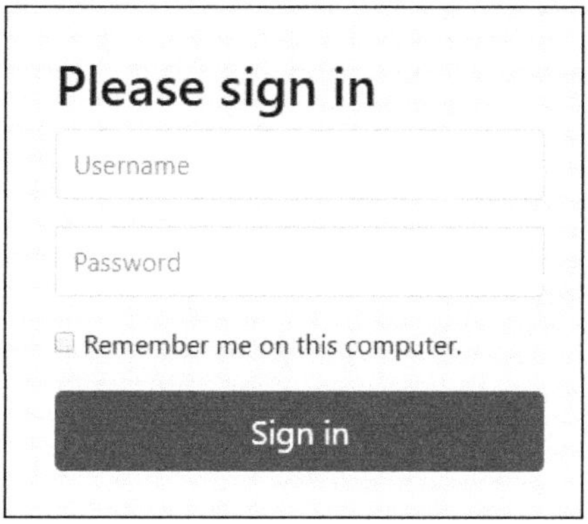

图 5-1

勾选“Remember me on this computer”可选框（简写为“remember-me”），按照正常的流程登录，并在开发者工具中查看浏览器 cookie，可以看到除 JSESSIONID 外多了一个值，如图 5-2 所示。

Name	Value	Do...	Path	Expires / Max-...	Size	H	Secure	Sa...
remember-me	YWRtaW46MTUyMjczN...	loc...	/	2018-04-03T06...	81	✓		
JSESSIONID	0E6737C0125EF884F6...	loc...	/	1969-12-31T23...	42	✓		

图 5-2

这就是 Spring Security 默认自动登录的 cookie 字段。在不配置的情况下，过期时间是两个星期。

```
public abstract class AbstractRememberMeServices implements RememberMeServices,
InitializingBean, LogoutHandler {
      public static final String SPRING_SECURITY_REMEMBER_ME_COOKIE_KEY =
"remember-me";
   public static final String DEFAULT_PARAMETER = "remember-me";
   public static final int TWO_WEEKS_S = 1209600;
   private static final String DELIMITER = ":";
   protected final Log logger = LogFactory.getLog(this.getClass());
   protected final MessageSourceAccessor messages =
SpringSecurityMessageSource.getAccessor();
   private UserDetailsService userDetailsService;
   private UserDetailsChecker userDetailsChecker = new
AccountStatusUserDetailsChecker();
   private AuthenticationDetailsSource<HttpServletRequest, ?>
authenticationDetailsSource = new WebAuthenticationDetailsSource();
   private String cookieName = "remember-me";
```

```
    private String cookieDomain;
    private String parameter = "remember-me";
    private boolean alwaysRemember;
    private String key;
    // 默认过期时间
    private int tokenValiditySeconds = 1209600;
    private Boolean useSecureCookie = null;
    private GrantedAuthoritiesMapper authoritiesMapper = new
NullAuthoritiesMapper();
    ...
}
```

Spring Security 会在每次表单登录成功之后更新此令牌，具体处理方式在源码中有体现。

```
// 散列加密部分
protected String makeTokenSignature(long tokenExpiryTime, String username,
String password) {
    String data = username + ":" + tokenExpiryTime + ":" + password + ":" +
this.getKey();

    MessageDigest digest;
    try {
        digest = MessageDigest.getInstance("MD5");
    } catch (NoSuchAlgorithmException var8) {
        throw new IllegalStateException("No MD5 algorithm available!");
    }

    return new String(Hex.encode(digest.digest(data.getBytes())));
}

...

public void onLoginSuccess(HttpServletRequest request, HttpServletResponse
response, Authentication successfulAuthentication) {
    String username = this.retrieveUserName(successfulAuthentication);
    String password = this.retrievePassword(successfulAuthentication);
    if (!StringUtils.hasLength(username)) {
        this.logger.debug("Unable to retrieve username");
    } else {
        if(!StringUtils.hasLength(password)) {
            UserDetails user =
this.getUserDetailsService().loadUserByUsername(username);
            password = user.getPassword();
            if (!StringUtils.hasLength(password)) {
```

```
            this.logger.debug("Unable to obtain password for user: " +
username);
            return;
        }
    }

    int tokenLifetime = this.calculateLoginLifetime(request,
successfulAuthentication);
    long expiryTime = System.currentTimeMillis();
    expiryTime += 1000L * (long)(tokenLifetime < 0?1209600:tokenLifetime);
    String signatureValue = this.makeTokenSignature(expiryTime, username,
password);
    // 登录成功，设置cookie
    this.setCookie(new String[]{username, Long.toString(expiryTime),
signatureValue}, tokenLifetime, request, response);
    if (this.logger.isDebugEnabled()) {
        this.logger.debug("Added remember-me cookie for user '" + username +
"', expiry: '" + new Date(expiryTime) + "'");
    }

    }
}
```

其中，在没有指定时，key 是一个 UUID 字符串。

```
private String getKey() {
    if(this.key == null) {
        this.key = UUID.randomUUID().toString();
    }

    return this.key;
}
```

这将导致每次重启服务后，key 都会重新生成，使得重启之前的所有自动登录 cookie 失效。除此之外，在多实例部署的情况下，由于实例间的 key 并不相同，所以当用户访问系统的另一个实例时，自动登录策略就会失效。合理的用法是指定 key。

```
@Autowired
private MyUserDetailsService userDetailsService;

@Override
protected void configure(HttpSecurity http) throws Exception {
    http
    .authorizeRequests()
```

```
            .antMatchers("/admin/api/**").hasAuthority("ROLE_ADMIN")
            .antMatchers("/user/api/**").hasRole("USER")
            .antMatchers("/app/api/**").permitAll()
            .anyRequest().authenticated()
            .and()
            .csrf().disable()
        .formLogin().permitAll()
            .and()
        .rememberMe()
            .userDetailsService(userDetailsService)
            .key("blurooo");
}
```

总体来说，这种方式不需要服务器花费空间来存储自动登录的相关数据，实现简单，安全性相对较高。但存在潜在风险，即如果该令牌在有效期内被盗取，那么用户的身份将完全暴露。

2. 持久化令牌方案

持久化令牌方案在交互上与散列加密方案一致，都是在用户勾选 Remember-me 之后，将生成的令牌发送到用户浏览器，并在用户下次访问系统时读取该令牌进行认证。不同的是，它采用了更加严谨的安全性设计。

在持久化令牌方案中，最核心的是 series 和 token 两个值，它们都是用 MD5 散列过的随机字符串。不同的是，series 仅在用户使用密码重新登录时更新，而 token 会在每一个新的 session 中都重新生成。

这样设计有什么好处呢？

首先，解决了散列加密方案中一个令牌可以同时在多端登录的问题。每个会话都会引发 token 的更新，即每个 token 仅支持单实例登录。

其次，自动登录不会导致 series 变更，而每次自动登录都需要同时验证 series 和 token 两个值，当该令牌还未使用过自动登录就被盗取时，系统会在非法用户验证通过后刷新 token 值，此时在合法用户的浏览器中，该 token 值已经失效。当合法用户使用自动登录时，由于该 series 对应的 token 不同，系统可以推断该令牌可能已被盗用，从而做一些处理。例如，清理该用户的所有自动登录令牌，并通知该用户可能已被盗号等。

在实现上，Spring Security 使用 PersistentRememberMeToken 来表明一个验证实体。

```java
public class PersistentRememberMeToken {
    private final String username;
    private final String series;
    private final String tokenValue;
    // 最后一次使用自动登录的时间
    private final Date date;

    ...
}
```

对应的，我们需要在数据库中新建一张 persistent_logins 表（存储自动登录信息的表），并建立如下结构（实际上，定制持久化登录信息表的意义不大，因为它对原系统几乎没有侵入性。而用户表不同，如果 Spring Security 不提供定制，就会导致原系统在接入时需要做大量更改）。

```sql
create table persistent_logins (
    username varchar(64) not null,
    series varchar(64) primary key,
    token varchar(64) not null,
    last_used timestamp not null
);
```

表的主键是 series，Spring Security 可以通过 series 查询表中的其他信息。有了存储自动登录信息的表之后，就可以继续配置 Spring Security 的 Remember-me 功能了。由于需要使用持久化令牌方案，所以定制 tokenRepository。

```java
@Autowired
private MyUserDetailsService userDetailsService;

@Override
protected void configure(HttpSecurity http) throws Exception {
    http
    .authorizeRequests()
        .antMatchers("/admin/api/**").hasAuthority("ROLE_ADMIN")
        .antMatchers("/user/api/**").hasRole("USER")
        .antMatchers("/app/api/**").permitAll()
        .anyRequest().authenticated()
        .and()
        .csrf().disable()
    .formLogin().permitAll()
        .and()
    .rememberMe()
        .userDetailsService(userDetailsService)
```

```
            .tokenRepository(null);
}
```

此处我们需要传入一个 PersistentTokenRepository 实例，PersistentTokenRepository 实例定义了持久化令牌的一些必要方法。

```
public interface PersistentTokenRepository {
    void createNewToken(PersistentRememberMeToken var1);

    void updateToken(String var1, String var2, Date var3);

    PersistentRememberMeToken getTokenForSeries(String var1);

    void removeUserTokens(String var1);
}
```

我们既可以按照自己的方式实现 PersistentTokenRepository 接口，也可以使用 Spring Security 提供的 JDBC 方案实现。

```
public class JdbcTokenRepositoryImpl extends JdbcDaoSupport implements PersistentTokenRepository {
    public static final String CREATE_TABLE_SQL = "create table persistent_logins (username varchar(64) not null, series varchar(64) primary key, token varchar(64) not null, last_used timestamp not null)";
    public static final String DEF_TOKEN_BY_SERIES_SQL = "select username,series,token,last_used from persistent_logins where series = ?";
    public static final String DEF_INSERT_TOKEN_SQL = "insert into persistent_logins (username, series, token, last_used) values(?,?,?,?)";
    public static final String DEF_UPDATE_TOKEN_SQL = "update persistent_logins set token = ?, last_used = ? where series = ?";
    public static final String DEF_REMOVE_USER_TOKENS_SQL = "delete from persistent_logins where username = ?";
    private String tokensBySeriesSql = "select username,series,token,last_used from persistent_logins where series = ?";
    private String insertTokenSql = "insert into persistent_logins (username, series, token, last_used) values(?,?,?,?)";
    private String updateTokenSql = "update persistent_logins set token = ?, last_used = ? where series = ?";
    private String removeUserTokensSql = "delete from persistent_logins where username = ?";
    ...
}
```

JdbcTokenRepositoryImpl 是基于 DataSource 实现对应 SQL 操作的类，所以我们需要为它指定 DataSource。

```java
@Autowired
private MyUserDetailsService userDetailsService;

@Autowired
private DataSource dataSource;

@Override
protected void configure(HttpSecurity http) throws Exception {
    JdbcTokenRepositoryImpl jdbcTokenRepository = new JdbcTokenRepositoryImpl();
    jdbcTokenRepository.setDataSource(dataSource);
    http
    .authorizeRequests()
        .antMatchers("/admin/api/**").hasAuthority("ROLE_ADMIN")
        .antMatchers("/user/api/**").hasRole("USER")
        .antMatchers("/app/api/**").permitAll()
        .anyRequest().authenticated()
        .and()
        .csrf().disable()
    .formLogin().permitAll()
        .and()
    .rememberMe()
        .userDetailsService(userDetailsService)
        .tokenRepository(jdbcTokenRepository);
}
```

与散列加密方案一样，持久化令牌方案也需要实现 UserDetailService。如果一切顺利那么重启服务后我们就已经引入持久化令牌的自动登录方案了。与散列加密方案相比，持久化令牌方案在用户体验层面没有任何变化。登录之后，可以看到 cookie 经过 Base64 解码之后不再是散列加密的形式了。

```
base64.decode(Z0drRzMyV3VlUkhKUTdzOHZPMGpNUT09OnRSY1Y0dUVRM3FDaHRvNm5JlGF8g=
c9PQ) = gGkG32WueRHJQ7s8vO0jMQ==:tRcV4uEQ3qChto6nIlGF8g==
```

冒号前的部分为 series，冒号后的部分为 token。当自动登录认证时，Spring Security 通过 series 获取用户名、token 以及上一次自动登录时间三个信息，通过用户名确认该令牌的身份，通过对比 token 获知该令牌是否有效，通过上一次自动登录时间获知该令牌是否已过期，并在完整校验通过之后生成新的 token。

```java
protected UserDetails processAutoLoginCookie(String[] cookieTokens,
HttpServletRequest request, HttpServletResponse response) {
    if(cookieTokens.length != 2) {
        throw new InvalidCookieException("Cookie token did not contain 2 tokens, but contained '" + Arrays.asList(cookieTokens) + "'");
    } else {
        String presentedSeries = cookieTokens[0];
        String presentedToken = cookieTokens[1];
        PersistentRememberMeToken token = this.tokenRepository.getTokenForSeries(presentedSeries);
        if(token == null) {
            throw new RememberMeAuthenticationException("No persistent token found for series id: " + presentedSeries);
        } else if(!presentedToken.equals(token.getTokenValue())) {
            this.tokenRepository.removeUserTokens(token.getUsername());
            throw new CookieTheftException(this.messages.getMessage("PersistentTokenBasedRememberMeServices.cookieStolen", "Invalid remember-me token (Series/token) mismatch. Implies previous cookie theft attack."));
        } else if(token.getDate().getTime() + (long)this.getTokenValiditySeconds() * 1000L < System.currentTimeMillis()) {
            throw new RememberMeAuthenticationException("Remember-me login has expired");
        } else {
            if(this.logger.isDebugEnabled()) {
                this.logger.debug("Refreshing persistent login token for user '" + token.getUsername() + "', series '" + token.getSeries() + "'");
            }

            PersistentRememberMeToken newToken = new PersistentRememberMeToken(token.getUsername(), token.getSeries(), this.generateTokenData(), new Date());

            try {
                this.tokenRepository.updateToken(newToken.getSeries(), newToken.getTokenValue(), newToken.getDate());
                this.addCookie(newToken, request, response);
            } catch (Exception var9) {
                this.logger.error("Failed to update token: ", var9);
                throw new RememberMeAuthenticationException("Autologin failed due to data access problem");
            }

            return
```

```
    this.getUserDetailsService().loadUserByUsername(token.getUsername());
        }
    }
}

protected void onLoginSuccess(HttpServletRequest request, HttpServletResponse
response, Authentication successfulAuthentication) {
    String username = successfulAuthentication.getName();
    this.logger.debug("Creating new persistent login for user " + username);
    PersistentRememberMeToken persistentToken = new
PersistentRememberMeToken(username, this.generateSeriesData(),
this.generateTokenData(), new Date());

    try {
        this.tokenRepository.createNewToken(persistentToken);
        this.addCookie(persistentToken, request, response);
    } catch (Exception var7) {
        this.logger.error("Failed to save persistent token ", var7);
    }

}
```

查看数据库 persistent_logins 表可以看到完整的信息，如图 5-3 所示。

```
mysql> select * from persistent_logins;
+----------+--------------------------+--------------------------+---------------------+
| username | series                   | token                    | last_used           |
+----------+--------------------------+--------------------------+---------------------+
| admin    | gGkG32WueRHJQ7s8v00jMQ== | tRcV4uEQ3qChto6nIlGF8g== | 2018-03-23 10:45:20 |
+----------+--------------------------+--------------------------+---------------------+
1 row in set (0.00 sec)
```

图 5-3

显然，两种方案都存在 cookie 被盗取导致身份被暂时利用的可能，如果有更高的安全性需求，建议使用 Spring Security 提供的令牌持久化方案。当然，最安全的方式还是尽量不使用自动登录，但很多时候，在实际开发中，优质体验比不可预期的安全风险要更为优先。

如果决定提供自动登录功能，就应当限制 cookie 登录时的部分执行权限。例如，修改密码、修改邮箱（防止找回密码）、查看隐私信息（如完整的手机号码、银行卡号等）等，校验登录密码或设置独立密码来做二次校验也是不错的方案。

5.3 注销登录

认证系统往往都带有注销登录功能，Spring Security 也提供了这方面的支持。事实上，从我们编写配置类继承 WebSecurityConfigurerAdapter 的那一刻起，Spring Security 就已经为我们的系统埋入了注销的逻辑。

```
protected final HttpSecurity getHttp() throws Exception {
    if (http != null) {
        return http;
    }

    DefaultAuthenticationEventPublisher eventPublisher = objectPostProcessor
            .postProcess(new DefaultAuthenticationEventPublisher());
localConfigureAuthenticationBldr.authenticationEventPublisher(eventPublisher
);

    AuthenticationManager authenticationManager = authenticationManager();
authenticationBuilder.parentAuthenticationManager(authenticationManager);
    authenticationBuilder.authenticationEventPublisher(eventPublisher);
    Map<Class<?>, Object> sharedObjects = createSharedObjects();

    http = new HttpSecurity(objectPostProcessor, authenticationBuilder,
            sharedObjects);
    if (!disableDefaults) {
        // @formatter:off
        http
                .csrf().and()
                .addFilter(new WebAsyncManagerIntegrationFilter())
                .exceptionHandling().and()
                .headers().and()
                .sessionManagement().and()
                .securityContext().and()
                .requestCache().and()
                .anonymous().and()
                .servletApi().and()
                .apply(new DefaultLoginPageConfigurer<>()).and()
                // 默认注册注销登录的过滤器
                .logout();
        // @formatter:on
        ClassLoader classLoader = this.context.getClassLoader();
```

```
        List<AbstractHttpConfigurer> defaultHttpConfigurers =
        SpringFactoriesLoader.loadFactories(AbstractHttpConfigurer.class,
classLoader);
        for (AbstractHttpConfigurer configurer : defaultHttpConfigurers) {
            http.apply(configurer);
        }
    }
    configure(http);
    return http;
}
```

HttpSecurity 内的 logout() 方法以一个 LogoutConfigurer 作为配置基础，创建一个用于注销登录的过滤器。

```
public void configure(H http) throws Exception {
   LogoutFilter logoutFilter = this.createLogoutFilter(http);
   http.addFilter(logoutFilter);
}

private LogoutFilter createLogoutFilter(H http) throws Exception {
   this.logoutHandlers.add(this.contextLogoutHandler);
   LogoutHandler[] handlers =
(LogoutHandler[])this.logoutHandlers.toArray(new
LogoutHandler[this.logoutHandlers.size()]);
   LogoutFilter result = new LogoutFilter(this.getLogoutSuccessHandler(),
handlers);
   result.setLogoutRequestMatcher(this.getLogoutRequestMatcher(http));
   result = (LogoutFilter)this.postProcess(result);
   return result;
}
```

它默认注册了一个 /logout 路由，用户通过访问该路由可以安全地注销其登录状态，包括使 HttpSession 失效、清空已配置的 Remember-me 验证，以及清空 SecurityContextHolder，并在注销成功之后重定向到 /login?logout 页面。

如有必要，还可以重新配置。

```
@Override
protected void configure(HttpSecurity http) throws Exception {
   http
      .logout()
```

```java
        // 指定接受注销请求的路由
        .logoutUrl("/myLogout")
        // 注销成功，重定向到该路径下
        .logoutSuccessUrl("/")
        // 注销成功的处理方式，不同于logoutSuccessUrl的重定向，logoutSuccessHandler
        // 更加灵活
        .logoutSuccessHandler(new LogoutSuccessHandler() {
            @Override
            public void onLogoutSuccess(HttpServletRequest httpServletRequest,
HttpServletResponse httpServletResponse, Authentication authentication) throws
IOException, ServletException {
                // 具体注销成功执行的逻辑
            }
        })
        // 使该用户的HttpSession失效
        .invalidateHttpSession(true)
        // 注销成功，删除指定的cookie
        .deleteCookies("cookie1", "cookie2")
        // 用于注销的处理句柄，允许自定义一些清理策略
        // 事实上LogoutSuccessHandler也能做到
        .addLogoutHandler(new LogoutHandler() {
            @Override
            public void logout(HttpServletRequest httpServletRequest,
HttpServletResponse httpServletResponse, Authentication authentication) {

            }
        });
}
```

实际上，logout 的清理过程是由多个 LogoutHandler 流式处理的。

```java
public interface LogoutHandler {
    void logout(HttpServletRequest var1, HttpServletResponse var2,
Authentication var3);
}
```

查看源码，在 Logout 过滤器中可以理顺整个注销的处理流程。

```java
public class LogoutFilter extends GenericFilterBean {
    private RequestMatcher logoutRequestMatcher;
    private final LogoutHandler handler;
    private final LogoutSuccessHandler logoutSuccessHandler;

    public LogoutFilter(LogoutSuccessHandler logoutSuccessHandler,
```

```java
LogoutHandler... handlers) {
    // 聚合所有的 LogoutHandler
    this.handler = new CompositeLogoutHandler(handlers);
    Assert.notNull(logoutSuccessHandler, "logoutSuccessHandler cannot be null");
    this.logoutSuccessHandler = logoutSuccessHandler;
    this.setFilterProcessesUrl("/logout");
}

public LogoutFilter(String logoutSuccessUrl, LogoutHandler... handlers) {
    this.handler = new CompositeLogoutHandler(handlers);
    Assert.isTrue(!StringUtils.hasLength(logoutSuccessUrl) || UrlUtils.isValidRedirectUrl(logoutSuccessUrl), logoutSuccessUrl + " isn't a valid redirect URL");
    SimpleUrlLogoutSuccessHandler urlLogoutSuccessHandler = new SimpleUrlLogoutSuccessHandler();
    if(StringUtils.hasText(logoutSuccessUrl)) {
        urlLogoutSuccessHandler.setDefaultTargetUrl(logoutSuccessUrl);
    }

    this.logoutSuccessHandler = urlLogoutSuccessHandler;
    this.setFilterProcessesUrl("/logout");
}

public void doFilter(ServletRequest req, ServletResponse res, FilterChain chain) throws IOException, ServletException {
    HttpServletRequest request = (HttpServletRequest)req;
    HttpServletResponse response = (HttpServletResponse)res;
    if(this.requiresLogout(request, response)) {
        Authentication auth = SecurityContextHolder.getContext().getAuthentication();
        if(this.logger.isDebugEnabled()) {
            this.logger.debug("Logging out user '" + auth + "' and transferring to logout destination");
        }
        // 执行聚合后的 logout 方法，实际上是迭代并执行这些 LogoutHandler
        this.handler.logout(request, response, auth);
        this.logoutSuccessHandler.onLogoutSuccess(request, response, auth);
    } else {
        chain.doFilter(request, response);
    }
}

...
```

```java
}

public final class CompositeLogoutHandler implements LogoutHandler {
    private final List<LogoutHandler> logoutHandlers;

    public CompositeLogoutHandler(LogoutHandler... logoutHandlers) {
        Assert.notEmpty(logoutHandlers, "LogoutHandlers are required");
        this.logoutHandlers = Arrays.asList(logoutHandlers);
    }

    public CompositeLogoutHandler(List<LogoutHandler> logoutHandlers) {
        Assert.notEmpty(logoutHandlers, "LogoutHandlers are required");
        this.logoutHandlers = logoutHandlers;
    }

    // 聚合后的 logout
    public void logout(HttpServletRequest request, HttpServletResponse response, Authentication authentication) {
        Iterator var4 = this.logoutHandlers.iterator();

        while(var4.hasNext()) {
            LogoutHandler handler = (LogoutHandler)var4.next();
            handler.logout(request, response, authentication);
        }

    }
}
```

第 6 章
会话管理

只需在两个浏览器中用同一个账号登录就会发现,到目前为止,系统尚未有任何会话并发限制。一个账户能多处同时登录可不是一个好的策略。事实上,Spring Security 已经为我们提供了完善的会话管理功能,包括会话固定攻击、会话超时检测以及会话并发控制。

6.1 理解会话

会话(session)就是无状态的 HTTP 实现用户状态可维持的一种解决方案。HTTP 本身的无状态使得用户在与服务器的交互过程中,每个请求之间都没有关联性。这意味着用户的访问没有身份记录,站点也无法为用户提供个性化的服务。session 的诞生解决了这个难题,服务器通过与用户约定每个请求都携带一个 id 类的信息,从而让不同请求之间有了关联,而 id 又可以很方便地绑定具体用户,所以我们可以把不同请求归类到同一用户。基于这个方案,为了让用户每个请求都携带同一个 id,在不妨碍体验的情况下,cookie 是很好的载体。当用户首次访问系统时,系统会为该用户生成一个 sessionId,并添加到 cookie 中。在该用户的会话期内,每个请求都自动携带该 cookie,因此系统可以很轻易地识别出这是来自哪个用户的请求。

尽管 cookie 非常有用,但有时用户会在浏览器中禁用它,可能是出于安全考虑,也可能是为了保护个人隐私。在这种情况下,基于 cookie 实现的 sessionId 自然就无法正常使用了。因此,有些服务还支持用 URL 重写的方式来实现类似的体验,例如:

```
http://blurooo.com;jsessionid=xxx
```

URL 重写原本是为了兼容禁用 cookie 的浏览器而设计的,但也容易被黑客利用。黑客只需访问一次系统,将系统生成的 sessionId 提取并拼凑在 URL 上,然后将该 URL 发给一些取得信任的用户。只要用户在 session 有效期内通过此 URL 进行登录,该 sessionId 就会绑定到用户的身份，黑客便可以轻松享有同样的会话状态，完全不需要用户名和密码，这就是典型的会话固定攻击。

6.2 防御会话固定攻击

防御会话固定攻击的方法非常简单，只需在用户登录之后重新生成新的 session 即可。在继承 WebSecurityConfigurerAdapter 时，Spring Security 已经启用了该配置。

```
protected final HttpSecurity getHttp() throws Exception {
    if (http != null) {
        return http;
    }

    DefaultAuthenticationEventPublisher eventPublisher = objectPostProcessor
            .postProcess(new DefaultAuthenticationEventPublisher());
localConfigureAuthenticationBldr.authenticationEventPublisher(eventPublisher);

    AuthenticationManager authenticationManager = authenticationManager();
authenticationBuilder.parentAuthenticationManager(authenticationManager);
    authenticationBuilder.authenticationEventPublisher(eventPublisher);
    Map<Class<?>, Object> sharedObjects = createSharedObjects();

    http = new HttpSecurity(objectPostProcessor, authenticationBuilder,
            sharedObjects);
    if (!disableDefaults) {
        // @formatter:off
        http
            .csrf().and()
            .addFilter(new WebAsyncManagerIntegrationFilter())
            .exceptionHandling().and()
            .headers().and()
            // 注册会话管理器
            .sessionManagement().and()
            .securityContext().and()
            .requestCache().and()
            .anonymous().and()
            .servletApi().and()
            .apply(new DefaultLoginPageConfigurer<>()).and()
```

```
            // 默认注册注销登录的过滤器
            .logout();
    // @formatter:on
    ClassLoader classLoader = this.context.getClassLoader();
    List<AbstractHttpConfigurer> defaultHttpConfigurers =
            SpringFactoriesLoader.loadFactories(AbstractHttpConfigurer.class, classLoader);

    for (AbstractHttpConfigurer configurer : defaultHttpConfigurers) {
        http.apply(configurer);
    }
}
configure(http);
return http;
}
```

sessionManagement 是一个会话管理的配置器,其中,防御会话固定攻击的策略有四种:

◎ none:不做任何变动,登录之后沿用旧的 session。
◎ newSession:登录之后创建一个新的 session。
◎ migrateSession:登录之后创建一个新的 session,并将旧的 session 中的数据复制过来。
◎ changeSessionId:不创建新的会话,而是使用由 Servlet 容器提供的会话固定保护。

默认已经启用 migrateSession 策略,如有必要,可以做出修改。

```
@Override
protected void configure(HttpSecurity http) throws Exception {
   http
   .authorizeRequests()
        .antMatchers("/admin/api/**").hasAuthority("ROLE_ADMIN")
        .antMatchers("/user/api/**").hasRole("USER")
        .antMatchers("/app/api/**").permitAll()
        .anyRequest().authenticated()
        .and()
        .csrf().disable()
     .formLogin().permitAll()
        .and()
     .sessionManagement()
        .sessionFixation().none();
}
```

在 Spring Security 中,即便没有配置,也大可不必担心会话固定攻击。这是因为 Spring

Security 的 HTTP 防火墙会帮助我们拦截不合法的 URL，当我们试图访问带 session 的 URL 时，实际上会被重定向到类似如图 6-1 所示的错误页。

图 6-1

具体细节可以翻看 Spring Security 源码，该部分内容在 StrictHttpFirewall 类中实现。

6.3 会话过期

除防御会话固定攻击外，还可以通过 Spring Security 配置一些会话过期策略。例如，会话过期时跳转到某个 URL。

```
.sessionManagement()
    .invalidSessionUrl("/session/invalid")
```

或者完全自定义过期策略。

```
// 实现 InvalidSessionStrategy 接口
public class MyInvalidSessionStrategy implements InvalidSessionStrategy {
    @Override
    public void onInvalidSessionDetected(HttpServletRequest httpServletRequest, HttpServletResponse httpServletResponse) throws IOException, ServletException
    {
        httpServletResponse.setContentType("application/json;charset=utf-8");
        httpServletResponse.getWriter().write("session 无效");
    }
}



@Override
protected void configure(HttpSecurity http) throws Exception {
    http
    .authorizeRequests()
        .antMatchers("/admin/api/**").hasAuthority("ROLE_ADMIN")
```

```
            .antMatchers("/user/api/**").hasRole("USER")
            .antMatchers("/app/api/**").permitAll()
            .anyRequest().authenticated()
            .and()
            .csrf().disable()
        .formLogin().permitAll()
            .and()
        .sessionManagement()
            // 配置 session 失效策略
            .invalidSessionStrategy(new MyInvalidSessionStrategy());
}
```

默认情况下，只要该会话在 30 分钟内没有活动便会失效，失效后再尝试发起访问，将会得到如图 6-2 所示的应答。

图 6-2

当然，我们可以手动修改会话的过期时间。

```
server.servlet.session.timeout = 60s
```

会话的过期时间最少为 1 分钟，所以即便设置小于 60 秒也会被修正为 1 分钟，这属于 Spring Boot 的配置策略。

```
private long getSessionTimeoutInMinutes() {
    long sessionTimeout = (long)this.getSessionTimeout();
    if(sessionTimeout > 0L) {
        sessionTimeout = Math.max(TimeUnit.SECONDS.toMinutes(sessionTimeout), 1L);
    }

    return sessionTimeout;
}
```

6.4　会话并发控制

固定会话攻击和会话过期策略都很简单，在 Spring Security 中，会话管理最完善的是会话并发控制，但会话并发控制存在一些用法陷阱，应当多加注意，下面来看看详细用法。

一个最简单的控制会话并发数的配置如下（为了避开陷阱，先启用基于内存的用户配置）。

```
@Bean
public UserDetailsService userDetailsService() {
    InMemoryUserDetailsManager manager = new InMemoryUserDetailsManager();
manager.createUser(User.withUsername("user").password("123").roles("USER").build());

manager.createUser(User.withUsername("admin").password("123").roles("USER", "ADMIN").build());
    return manager;
}

@Override
protected void configure(HttpSecurity http) throws Exception {
    http
    .authorizeRequests()
        .antMatchers("/admin/api/**").hasAuthority("ROLE_ADMIN")
        .antMatchers("/user/api/**").hasRole("USER")
        .antMatchers("/app/api/**").permitAll()
        .anyRequest().authenticated()
        .and()
    .formLogin().permitAll()
        .and()
    .sessionManagement()
        // 最大会话数设置为 1
        .maximumSessions(1);
}
```

maximumSessions 用于设置单个用户允许同时在线的最大会话数，如果没有额外配置，那么新登录的会话会踢掉旧的会话，如图 6-3 所示。

图 6-3

具体的实现细节在 ConcurrentSessionControlAuthenticationStrategy 类中可以看到。

```
public class ConcurrentSessionControlAuthenticationStrategy implements
MessageSourceAware, SessionAuthenticationStrategy {
```

```java
    protected MessageSourceAccessor messages =
SpringSecurityMessageSource.getAccessor();
    private final SessionRegistry sessionRegistry;
    private boolean exceptionIfMaximumExceeded = false;
    private int maximumSessions = 1;

    ...

    public void onAuthentication(Authentication authentication,
HttpServletRequest request, HttpServletResponse response) {
        List<SessionInformation> sessions =
this.sessionRegistry.getAllSessions(authentication.getPrincipal(), false);
        int sessionCount = sessions.size();
        int allowedSessions =
this.getMaximumSessionsForThisUser(authentication);
        if(sessionCount >= allowedSessions) {
            if(allowedSessions != -1) {
                // 当已存在的会话数等于最大会话数时
                if(sessionCount == allowedSessions) {
                    HttpSession session = request.getSession(false);
                    if(session != null) {
                        Iterator var8 = sessions.iterator();

                        while(var8.hasNext()) {
                            SessionInformation si =
(SessionInformation)var8.next();
                            // 当前验证的会话如果并非新的会话，则不做任何处理
                            if(si.getSessionId().equals(session.getId())) {
                                return;
                            }
                        }
                    }
                }

                // 否则进行策略判断
                this.allowableSessionsExceeded(sessions, allowedSessions,
this.sessionRegistry);
            }
        }
    }

    protected void allowableSessionsExceeded(List<SessionInformation> sessions,
int allowableSessions, SessionRegistry registry) throws
```

```
SessionAuthenticationException {
    // exceptionIfMaximumExceeded 指示了当用户达到最大会话数时,是否阻止新会话建立
    if(!this.exceptionIfMaximumExceeded && sessions != null) {
        SessionInformation leastRecentlyUsed = null;
        Iterator var5 = sessions.iterator();

        while(true) {
            SessionInformation session;
            do {
                if(!var5.hasNext()) {
                    // 当新会话建立时,使最早的会话过期
                    leastRecentlyUsed.expireNow();
                    return;
                }

                session = (SessionInformation)var5.next();
            } while(leastRecentlyUsed != null
&& !session.getLastRequest().before(leastRecentlyUsed.getLastRequest()));

            leastRecentlyUsed = session;
        }
    } else {
        throw new
SessionAuthenticationException(this.messages.getMessage("ConcurrentSessionCo
ntrolAuthenticationStrategy.exceededAllowed", new
Object[]{Integer.valueOf(allowableSessions)}, "Maximum sessions of {0} for this
principal exceeded"));
    }
}
...
}
```

如果我们需要在会话数达到最大数时,阻止新会话建立,而不是踢掉旧的会话,则可以像下面这样配置。

```
.sessionManagement()
   .maximumSessions(1)
   // 阻止新会话登录,默认为 false
   .maxSessionsPreventsLogin(true);
```

实际运行之后貌似没有问题,当建立新会话时,确实被阻止了,如图 6-4 所示。

如果此时庆幸 Spring Security 的会话并发控制如此简单,那未免高兴得过早。怎么回事呢?

我们首先尝试将已登录的旧会话注销（通常是访问/logout），理论上应该可以继续登录了，但很遗憾，Spring Security 依然提示我们超过了最大会话数。事实上，除非重启服务，否则该用户将很难再次登录系统。这是因为 Spring Security 是通过监听 session 的销毁事件来触发会话信息表相关清理工作的，但我们并没有注册过相关的监听器，导致 Spring Security 无法正常清理过期或已注销的会话。

图 6-4

在 Servlet 中，监听 session 相关事件的方法是实现 HttpSessionListener 接口，并在系统中注册该监听器。Spring Security 在 HttpSessionEventPublisher 类中实现 HttpSessionEventPublisher 接口，并转化成 Spring 的事件机制。

```java
public class HttpSessionEventPublisher implements HttpSessionListener {
    private static final String LOGGER_NAME =
HttpSessionEventPublisher.class.getName();

    public HttpSessionEventPublisher() {
    }

    ApplicationContext getContext(ServletContext servletContext) {
        return
SecurityWebApplicationContextUtils.findRequiredWebApplicationContext(servletContext);
    }

    public void sessionCreated(HttpSessionEvent event) {
        HttpSessionCreatedEvent e = new
HttpSessionCreatedEvent(event.getSession());
        Log log = LogFactory.getLog(LOGGER_NAME);
        if(log.isDebugEnabled()) {
            log.debug("Publishing event: " + e);
        }

        // 调用 ApplicationEventPublisher，重新发布 session 创建事件
```

```
this.getContext(event.getSession().getServletContext()).publishEvent(e);
    }

    public void sessionDestroyed(HttpSessionEvent event) {
        HttpSessionDestroyedEvent e = new
HttpSessionDestroyedEvent(event.getSession());
        Log log = LogFactory.getLog(LOGGER_NAME);
        if(log.isDebugEnabled()) {
            log.debug("Publishing event: " + e);
        }

        // 调用 ApplicationEventPublisher, 重新发布 session 销毁事件
this.getContext(event.getSession().getServletContext()).publishEvent(e);
    }
}
```

在 Spring 事件机制中，事件的发布、订阅都交由 Spring 容器来托管，我们可以很方便地通过注册 bean 的方式来订阅关心的事件。

显然，前面的配置无法正常工作是因为缺少了事件源（会话清理订阅的 Spring 事件，而非原生的 HttpSessionEvent），另外，还需要把 HttpSessionEventPublisher 注册到 IoC 容器中，才能将 Java 事件转化成 Spring 事件。

```
@Bean
public HttpSessionEventPublisher httpSessionEventPublisher() {
    return new HttpSessionEventPublisher();
}
```

这个问题在新会话剔除旧会话的策略中就已存在，只是没有明显暴露出来，因此只要我们使用会话管理功能，就应该同时配置 HttpSessionEventPublisher。

至此，我们已经基本掌握了如何配置 Spring Security 的会话管理，但在真实场景中，可能还会碰到一些陷阱。前面提到过，为了避开陷阱，可以先启用基于内存的用户配置。陷阱是什么？为什么基于内存的用户配置就可以避开呢？

我们沿用前面章节自定义数据库结构认证的方式，先实现 UserDetails。

```
// 继承 UserDetails 的用户对象, 即 Spring Security 中的 principal
public class User implements UserDetails {
```

```java
    private Long id;

    private String username;

    private String password;

    private String roles;

    private boolean enable;

    private List<GrantedAuthority> authorities;

    public String getRoles() {
        return roles;
    }

    public void setRoles(String roles) {
        this.roles = roles;
    }

    public Long getId() {
        return id;
    }

    public void setId(Long id) {
        this.id = id;
    }

    public boolean isEnable() {
        return enable;
    }

    public void setEnable(boolean enable) {
        this.enable = enable;
    }

    public String getUsername() {
        return username;
    }

    public void setUsername(String username) {
        this.username = username;
    }
```

```java
    @Override
    public boolean isAccountNonExpired() {
        return true;
    }

    @Override
    public boolean isAccountNonLocked() {
        return true;
    }

    @Override
    public boolean isCredentialsNonExpired() {
        return true;
    }

    @Override
    public boolean isEnabled() {
        return this.enable;
    }

    public void setAuthorities(List<GrantedAuthority> authorities) {
        this.authorities = authorities;
    }

    @Override
    public Collection<? extends GrantedAuthority> getAuthorities() {
        return this.authorities;
    }

    public String getPassword() {
        return password;
    }

    public void setPassword(String password) {
        this.password = password;
    }
}
```

接着实现 UserDetailsService。

```java
@Service
public class MyUserDetailsService implements UserDetailsService {

    @Autowired
```

```java
    private UserService userService;

    @Override
    public UserDetails loadUserByUsername(String username) throws
UsernameNotFoundException {
        User user = userService.findByUserName(username);
        if (user == null) {
            throw new UsernameNotFoundException("用户不存在");
        }
        user.setAuthorities(generateAuthorities(user.getRoles()));
        return user;
    }

    private List<GrantedAuthority> generateAuthorities(String roles) {
        List<GrantedAuthority> authorities = new ArrayList<>();
        if (roles != null && !"".equals(roles)) {
            for (String role : roles.split(",")) {
                authorities.add(new SimpleGrantedAuthority(role));
            }
        }
        return authorities;
    }
}
```

把基于内存的用户配置去掉，确保仅有一个 UserDetailsService 对象可注入。不出意外，我们已经将验证的数据源切换到了数据库。登录功能正常，但会话并发控制似乎并不起作用，这是怎么回事呢？要想理解这个问题，还需要深入了解 Spring Security 关于会话管理的设计方式。

Spring Security 为了实现会话并发控制，采用会话信息表来管理用户的会话状态，具体实现见 SessionRegistryImpl 类。

```java
// 实现ApplicationListener接口，可以监听Spring事件
public class SessionRegistryImpl implements SessionRegistry,
ApplicationListener<SessionDestroyedEvent> {
    protected final Log logger = LogFactory.getLog(SessionRegistryImpl.class);
    // 存放用户以及其对应的所有sessionId的map
    private final ConcurrentMap<Object, Set<String>> principals;
    // 存放sessionId以及其对应的SessionInformation
    private final Map<String, SessionInformation> sessionIds;

    public SessionRegistryImpl() {
        this.principals = new ConcurrentHashMap();
        this.sessionIds = new ConcurrentHashMap();
```

```java
    }

    public SessionRegistryImpl(ConcurrentMap<Object, Set<String>> principals,
Map<String, SessionInformation> sessionIds) {
        this.principals = principals;
        this.sessionIds = sessionIds;
    }

    public List<Object> getAllPrincipals() {
        return new ArrayList(this.principals.keySet());
    }

    public List<SessionInformation> getAllSessions(Object principal, boolean
includeExpiredSessions) {
        Set<String> sessionsUsedByPrincipal =
(Set)this.principals.get(principal);
        if(sessionsUsedByPrincipal == null) {
            return Collections.emptyList();
        } else {
            List<SessionInformation> list = new
ArrayList(sessionsUsedByPrincipal.size());
            Iterator var5 = sessionsUsedByPrincipal.iterator();

            while(true) {
                SessionInformation sessionInformation;
                do {
                    do {
                        if(!var5.hasNext()) {
                            return list;
                        }

                        String sessionId = (String)var5.next();
                        sessionInformation =
this.getSessionInformation(sessionId);
                    } while(sessionInformation == null);
                } while(!includeExpiredSessions &&
sessionInformation.isExpired());

                list.add(sessionInformation);
            }
        }
    }

    public SessionInformation getSessionInformation(String sessionId) {
```

```java
        Assert.hasText(sessionId, "SessionId required as per interface
contract");
        return (SessionInformation)this.sessionIds.get(sessionId);
    }

    // 实现 onApplicationEvent 接口，表明处理 SessionDestroyedEvent 事件
    public void onApplicationEvent(SessionDestroyedEvent event) {
        String sessionId = event.getId();
        // 当会话销毁事件被触发时，移除对应 sessionId 的相关数据
        this.removeSessionInformation(sessionId);
    }

    public void refreshLastRequest(String sessionId) {
        Assert.hasText(sessionId, "SessionId required as per interface
contract");
        SessionInformation info = this.getSessionInformation(sessionId);
        if(info != null) {
            info.refreshLastRequest();
        }

    }

    // 注册新的会话
    // SessionManagementConfigure 默认会将 RegisterSessionAuthenticationStrategy
    // 添加到一个组合式的 SessionAuthenticationStrategy 中，并由
    // AbstractAuthenticationProcessingFilter 在登录成功时调用，从而触发
    // registerNewSession 动作
    public void registerNewSession(String sessionId, Object principal) {
        Assert.hasText(sessionId, "SessionId required as per interface
contract");
        Assert.notNull(principal, "Principal required as per interface
contract");
        if(this.logger.isDebugEnabled()) {
            this.logger.debug("Registering session " + sessionId + ", for
principal " + principal);
        }

        if(this.getSessionInformation(sessionId) != null) {
            this.removeSessionInformation(sessionId);
        }

        this.sessionIds.put(sessionId, new SessionInformation(principal,
sessionId, new Date()));
```

```java
    // 判断该用户是否已经存在
    Set<String> sessionsUsedByPrincipal = 
(Set)this.principals.get(principal);
    if(sessionsUsedByPrincipal == null) {
        sessionsUsedByPrincipal = new CopyOnWriteArraySet();
        // 若不存在,则添加
        Set<String> prevSessionsUsedByPrincipal = 
(Set)this.principals.putIfAbsent(principal, sessionsUsedByPrincipal);
        if(prevSessionsUsedByPrincipal != null) {
            sessionsUsedByPrincipal = prevSessionsUsedByPrincipal;
        }
    }

    ((Set)sessionsUsedByPrincipal).add(sessionId);
    if(this.logger.isTraceEnabled()) {
        this.logger.trace("Sessions used by '" + principal + "' : " + 
sessionsUsedByPrincipal);
    }

}

// 移除对应的会话信息
public void removeSessionInformation(String sessionId) {
    Assert.hasText(sessionId, "SessionId required as per interface contract");
    SessionInformation info = this.getSessionInformation(sessionId);
    if(info != null) {
        if(this.logger.isTraceEnabled()) {
            this.logger.debug("Removing session " + sessionId + " from set of registered sessions");
        }

        // 以 String 类型的 key 移除对应的 sessionId
        this.sessionIds.remove(sessionId);
        // 以 Object 类型的 key 获取对应用户的所有 sessionId
        Set<String> sessionsUsedByPrincipal = 
(Set)this.principals.get(info.getPrincipal());
        // 如果获取成功,则做对应的清理工作
        if(sessionsUsedByPrincipal != null) {
            if(this.logger.isDebugEnabled()) {
                this.logger.debug("Removing session " + sessionId + " from principal's set of registered sessions");
```

```
                }
            sessionsUsedByPrincipal.remove(sessionId);
            if(sessionsUsedByPrincipal.isEmpty()) {
                if(this.logger.isDebugEnabled()) {
                    this.logger.debug("Removing principal " +
info.getPrincipal() + " from registry");
                }

                this.principals.remove(info.getPrincipal());
            }

            if(this.logger.isTraceEnabled()) {
                this.logger.trace("Sessions used by '" + info.getPrincipal()
+ "' : " + sessionsUsedByPrincipal);
            }
        }
    }
}
```

为了更直观地认识 principals 和 sessionIds 的实际结构，可以参照图 6-5 所示的实际运行期来存储数据。

图 6-5

值得一提的是，principals 采用了以用户信息为 key 的设计。我们知道，在 hashMap 中，以对象为 key 必须覆写 hashCode 和 equals 两个方法（具体原因可以查阅 hashMap 的设计），但我们实现 UserDetails 时并没有这么做，这导致同一个用户每次登录注销时计算得到的 key 都不相同，所以每次登录都会向 principals 中添加一个用户，而注销时却从来不能有效移除。在这种情况下，不仅达不到会话并发控制的效果，还会引发内存泄露。

理解了缘由之后，要避开陷阱自然不难，为自定义的用户类覆写 hashCode 和 equals 两个方法即可。

```
@Override
public boolean equals(Object obj) {
    return obj instanceof User ? this.username.equals(((User)obj).username) : false;
}

@Override
public int hashCode() {
    return this.username.hashCode();
}
```

为什么基于内存的用户配置不会触发陷阱，主要也是基于这一点。因为在几乎所有我们可以见到的会话管理配置示例中，包括官方示例，它们通常是可以正常工作的，示例程序很少会提供相对复杂的自定义 UserDetails 方案，大多使用简单的基于内存的用户配置，这些都沿用了 Spring Security 内部实现的 UserDetails，自然不会出现问题。

```
public class User implements UserDetails, CredentialsContainer {
    private static final long serialVersionUID = 420L;
    private String password;
    private final String username;
    private final Set<GrantedAuthority> authorities;
    private final boolean accountNonExpired;
    private final boolean accountNonLocked;
    private final boolean credentialsNonExpired;
    private final boolean enabled;
...

    public boolean equals(Object rhs) {
        return rhs instanceof User?this.username.equals(((User)rhs).username):false;
    }
```

```
public int hashCode() {
    return this.username.hashCode();
}
```
}

6.5　集群会话的缺陷

会话通常保存在服务器内存中，客户端访问时根据自己的 sessionId 在内存中查找，这种方法虽然简单快捷，但缺点也很明显。从容量上来说，服务器内存有限，除了系统正常运行的消耗，留给 session 的空间不多，当访问量增大时，内存就会捉襟见肘。从稳定性上来说，session 依赖于内存，而内存并非持久性存储容器，就算服务器本身是可靠的，但当部署在上面的服务停止或重启时，也会导致所有会话状态丢失。当然，这两个缺点还只是体验性缺陷，并不足以影响可用性，在单机部署时为了节省精力忽略这两个问题也是可以的。但当我们的系统采用集群部署时，就会有更多关于可用性的问题需要考虑。

大部分的集群部署会采用类似图 6-6 所示的网络结构。

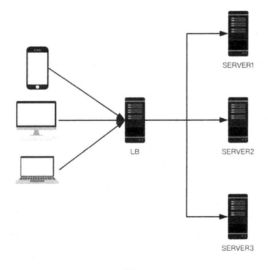

图 6-6

在这种网络结构下，用户的请求首先会打在 LB 服务器上（Load Balance，即负载均衡，常见的有 Nginx、HAProxy 等），LB 服务器再根据负载策略将这些请求转发至后面的服务，以达到请求分散的目的。正常来说，在集群环境下，同个用户的请求可能会被分发到不同的服务器上，假如登录操作是在 SERVER1 完成的，即 SERVER1 缓存了用户的登录状态，但 SERVER2 和 SERVER3 并不知情，如果该用户的后续操作被分配到了 SERVER2 或 SERVER3 上，这时就会要求该用户重新登录，这就是典型的会话状态集群不同步问题。

6.6 集群会话的解决方案

解决集群会话的常见方案有三种：

◎ session 保持
◎ session 复制
◎ session 共享

session 保持也叫粘滞会话（Sticky Sessions），通常采用 IP 哈希负载策略将来自相同客户端的请求转发至相同的服务器上进行处理。session 保持虽然避开了集群会话，但也存在一些缺陷。例如，某个营业部的网络使用同个 IP 出口，那么使用该营业部网络的所有员工实际的源 IP 其实是同一个，在 IP 哈希负载策略下，这些员工的请求都将被转发到相同的服务器上，存在一定程度的负载失衡。

session 复制是指在集群服务器之间同步 session 数据，以达到各个实例之间会话状态一致的做法。但毫无疑问，在集群服务器之间进行数据同步的做法非常不可取，尤其是在服务器实例很多的情况下，任何变动都需要其他所有实例同步，不仅消耗数据带宽，还会占用大量的资源。

相较于前两种方案，session 共享则要实用得多。session 共享是指将 session 从服务器内存抽离出来，集中存储到独立的数据容器，并由各个服务器共享，如图 6-7 所示。

由于所有的服务器实例单点存取 session，所以集群不同步的问题自然也就不存在了，而且独立的数据容器容量相较于服务器内存要大得多。另外，与服务本身分离、可持久化等特性使得会话状态不会因为服务停止而丢失。当然，session 共享并非没有缺点，独立的数据容器增加了网络交互，数据容器的读/写性能、稳定性以及网络 I/O 速度都成为性能的瓶颈。基于这些问题，尽管在理论上使用任何存储介质都可以实现 session 共享，但在内网环境下，高可用部署的 Redis 服务器无疑为最优选择。Redis 基于内存的特性让它拥有极高的读/写性能，高可用部署不仅降低了网络 I/O 损耗，还提高了稳定性。

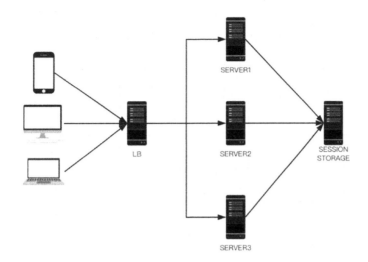

图 6-7

6.7 整合 Spring Session 解决集群会话问题

Spring Security 提供的会话并发控制是基于内存实现的，在集群部署时如果想要使用会话并发控制，则必须进行适配。

session 共享，本质上就是存储容器的变动，但如何得到最优存取结构、如何准确清理过期会话，以及如何整合 WebSocket 等无法回避。Spring Session 就是专门用于解决集群会话问题的，它不仅为集群会话提供了非常完善的支持，与 Spring Security 的整合也有专门的实现。

Spring Session 支持多种类型的存储容器，包括 Redis、MongoDB 等。由于接下来的整合都是基于 Redis 的，所以读者可以自行准备 Redis 测试环境，具体如何安装不再赘述。

在 pom.xml 中引入三个依赖。

```
<!--spring session 核心依赖-->
<dependency>
        <groupId>org.springframework.session</groupId>
        <artifactId>spring-session-core</artifactId>
</dependency>
<!--spring session 对接 Redis 的必要依赖-->
<dependency>
        <groupId>org.springframework.session</groupId>
```

```xml
        <artifactId>spring-session-data-redis</artifactId>
</dependency>
<!--spring boot 整合的 Redis 核心依赖-->
<dependency>
        <groupId>org.springframework.boot</groupId>
        <artifactId>spring-boot-starter-data-redis</artifactId>
</dependency>
```

之后就可以配置 Spring Session 了，主要是为 Spring Security 提供集群支持的会话注册表。

```java
// 启用基于 Redis 的 httpSession 实现
@EnableRedisHttpSession
public class HttpSessionConfig {

    // 提供 Redis 连接，默认是 localhost:6379
    @Bean
    public RedisConnectionFactory connectionFactory() {
        return new JedisConnectionFactory();
    }

    @Autowired
    private FindByIndexNameSessionRepository sessionRepository;

    // SpringSessionBackedSessionRegistry 是 session 为 Spring Security 提供的
    // 用于在集群环境下控制会话并发的会话注册表实现类
    @Bean
    public SpringSessionBackedSessionRegistry sessionRegistry() {
        return new SpringSessionBackedSessionRegistry(sessionRepository);
    }

    // httpSession 的事件监听，改用 session 提供的会话注册表
    @Bean
    public HttpSessionEventPublisher httpSessionEventPublisher() {
        return new HttpSessionEventPublisher();
    }
}
```

最后将新的会话注册表提供给 Spring Security 即可。

```java
@Autowired
private SpringSessionBackedSessionRegistry redisSessionRegistry;

@Override
protected void configure(HttpSecurity http) throws Exception {
```

```
http
.authorizeRequests()
    .antMatchers("/admin/api/**").hasAuthority("ROLE_ADMIN")
    .antMatchers("/user/api/**").hasRole("USER")
    .antMatchers("/app/api/**").permitAll()
    .anyRequest().authenticated()
    .and()
    .csrf().disable()
.formLogin().permitAll()
    .and()
.sessionManagement()
    .maximumSessions(1)
    // 使用session提供的会话注册表
    .sessionRegistry(redisSessionRegistry);
}
```

Spring Session 与 Spring Security 的整合非常简单，如果要验证是否生效，则可以在不同端口上启动多个实例，并在各自端口实例上登录同个账号。若出现挤占，则代表配置生效。登录成功之后，Redis 会存储用户 session 相关的数据，默认命名为 spring:session，如图 6-8 所示。

```
my:0>keys spring:session:*
1) spring:session:sessions:expires:e466f5f6-6c48-43ef-8571-79d2ff53ee71
2) spring:session:expirations:1523271420000
3) spring:session:index:org.springframework.session.FindByIndexNameSessionRepository.PRINCIPAL_NAME_INDEX_NAME:admin
4) spring:session:sessions:e466f5f6-6c48-43ef-8571-79d2ff53ee71
```

图 6-8

Spring Session 还提供了与 Spring Security 自动登录的整合，感兴趣的读者可以自行了解。

第 7 章
密码加密

密码安全是互联网安全的一个缩影，我们在享受互联网服务的同时，也应当对它投入更多的关注。

7.1 密码安全的重要性

2011 年 12 月，国内某开发者社区网站被黑客"拖库"，600 多万个密码明文存储的用户账号被公开，大量用户直接面临数据隐私泄露和数据安全的双重威胁。

这次事件为我们敲响了警钟，一旦发生"拖库"，如何尽可能地减少用户的损失，在每一个系统中都不应被忽略。从开发者的角度而言，可以落实在如何安全存储用户密码上。

为什么要安全存储用户密码呢？当被"拖库"时，当前系统的所有数据就成为既定损失，如果用户的密码同时被泄露，且未做加密，那么即便及时修复了"拖库"的问题，后续也可能会导致扩散性损失。因为黑客手里还握着大量的用户密码，安全威胁随时都在，这时能做的只能是紧急升级系统，强制所有用户验证手机号码或邮箱，并重置密码方能再次进入，这显然很被动。如果在系统开发之初就有安全存储用户密码的意识，则能有效防止这种情况。

7.2 密码加密的演进

安全存储用户密码很多时候都是使用加密的方式来进行的。提到加密，诸如 MD5、SHA 等摘要算法通常会第一时间浮现在我们脑海里，尽管它们并非真正的加密（不可逆）。例如，在密码加密这个场景下，对用户密码进行 MD5 运算后再存储就是一种看起来可行的思路，只要在每次登录验证时，对用户提交的密码进行同样的 MD5 运算，然后与数据库中存储的值进行对比即可。

事实上，尽管摘要算法无法通过逆运算获得原文，但由于摘要值有限（MD5 算法最多只能表示 36 的 16 次方个摘要值），而原文无限，所以一定存在两个甚至更多个不同数据通过运算得到同一摘要值的情况，即存在被"碰撞"的可能。

除"碰撞"外，最直接的破解方式是构建反查表。相同串（密码）经过 MD5 运算后一定会得到同一个值，于是有些人便制作了这样一张表，如表 7-1 所示。

表 7-1

49ba59abbe56e057	123456
7a57a5a743894aoe	admin
8458ce06fbc5bb76	qwerty
16e0c197e42a6be3	5201314
……	……

这张表穷举了几乎所有的常用密码，专门用于实现 MD5（其他散列算法同理）的快速反查（MD5 分为 32 位和 16 位，事实上，除了长度没有其他区别。将 32 位的前 8 位和后 8 位去掉即可得到 16 位，生成表时自然会选择位数少的，这样空间占用小）。如果我们的密码恰好被反查表收录，那么黑客几乎不费吹灰之力就可以直接完成破解。可想而知，这种破解方式的成功率完全取决于反查表的覆盖面是否足够大，以及数据是否足够精准。熟悉暴力破解的读者可能知道，一个强大的"字典"对破解的成功率和效率有非常大的帮助，反查表在一定程度上类似于"字典"。如果"字典"被穷举完了还找不到密码，就意味着破解失败。

反查表是一种初级的破解手段，只能应付一些可以预想到的密码形式。理论上只要我们的密码设置得足够长，字符组成足够复杂，就很难被收录。反查表的大小受限于储存设备的容量大小（内存或硬盘），因而通常只列举一些常见的密码。

基于反查表的缺陷，在无法拥有更多内存的情况下，如何增加表的容量是一个值得思考的方向。瑞士洛桑联邦技术学院的 Philippe Oechslin 提出了一种优化型的时间—内存交换算法，让表不再只是存储明文与密文的一一对应关系，而是助力于穷举计算的一个查找数据集，从而允许在有限的内存中以极高的效率破解更多可能的密码。受该思路的启发，一些黑客改进了部分流程，形成了大名鼎鼎的彩虹表。

下面简单介绍彩虹表的原理，如图 7-1 所示。

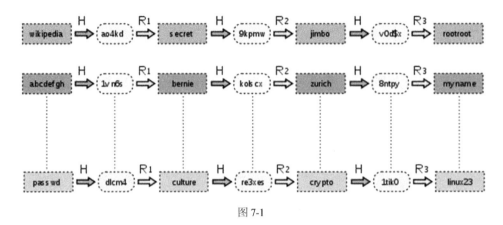

图 7-1

彩虹表的大小通常超过 100GB，不同于反查表，彩虹表没有直接存储摘要值与原文的一一对应关系，而是存储了一个时间—空间平衡的散列链集。可以认为每一条散列链代表了一组相同特征的明文，但每一条散列链并不需要完整存储，仅仅存储起始节点和末节点即可。例如，在图 7-1 中，H 代表一次散列计算，R_1、R_2、R_3 代表一次递归性质的约减计算，R_3 中的 3 指定了一条散列链会经历 3 次约减，即代表了这条散列链可以携带 3 条明文（后面用 K 指代）。当需要破解一个散列值时，只要对其做相对应的约减计算，便能推导出其所处的散列链，再由散列链正推就可能直接得到该散列值的明文。彩虹表相比于反查表，由于 K 的存在，使得每一条记录都能携带更多的明文，即减少了空间占用。但也由于 K 的存在，在进行破解时，需要付出相应的运算量（关于 K 的等差数列求和）。实际上，在普通计算机上辅以 CUDA 技术（显卡厂商 NVIDIA 推出的基于 GPU 的运算平台），可以达到每秒千亿次的破解速度，这就使得散列加密很多时候只是充当了遮羞布的作用。

尽管在彩虹表面前散列加密是脆弱的，但也并非束手无策，因为彩虹表也有弱点，加盐是最简单有效地防御彩虹表的手段。

所谓加盐加密，是指在计算摘要值之前，为原文附上额外的随机值，以达到扰乱目的的加密方式。具体实现方法并不固定。例如，以用户名作为盐值。

```
username = "blurooo"
password = "security"
persistentPwd = md5(username + ":" + password)
```

为了使盐值真正随机，可以将盐值作为用户的一个数据字段存储，并用 UUID 之类的字符串来表示。

```
salt = UUID.randomUUID()
password = "security"
persistentPwd = md5(password + ":" + salt)
```

或者将随机盐值直接挂在加密后的摘要值上，省去额外的存储字段。

```
salt = UUID.randomUUID()
password = "security"
persistentPwd = salt + ":" + md5(password + ":" + salt)
```

反查表和彩虹表都是通过对散列值或散列值的加工从而逆向推导出密码的，加盐等同于阻断了散列值与密码的直接对应关系，使得逆向推导的破解方式不再具有威慑力，但最经典的穷举法依然会带来威胁。随机盐值通常在实际存储密码时会同步存储，即当发生"拖库"时，盐值也很难幸免。在正向穷举时，盐值的阻碍非常小。例如，穷举 1 至 8 位的密码，只需相应附上盐值即可，不用在乎该盐值有多长，而且其计算耗时在当前来说是完全可以接受的，尤其在算力强的计算平台上（计算能力随着时间的推移会越来越强，意味着暴力破解当前的散列密码会越来越容易）。

既然穷举法如此简单粗暴，那么有制约的手段吗？当然有，"蛇打七寸"，穷举破解十分依赖于计算能力，常规的散列算法因为速度快反而助推了穷举的速度，所以我们需要一种慢的加密手段，慢到暴力破解无法忍受，慢到超级计算机也无可奈何，BCrypt 正是这样一种算法。

"blurooo"的 BCrypt 密文如下所示：

`$2a$12$dw8zzxcArWCIucx.fOi7/O.Ky2eX/R6jTagF/6Kcly/jeQe/H0tE6`

其中，2a 表明了算法的版本。2a 版本加入了对非 ASCII 字符以及空终止符的处理，类似的版本号还有 2x、2y、2b 等，都是在前面版本的基础上修复了某些缺陷或新增了某些特性。12 是一个成本参数，它表明该密文需要迭代的次数。12 是指 2 的 12 次方，即 4096 次。BCrypt 依靠此参数来限制算法的速度。成本参数的理想取值是既让暴力破解无法忍受，又不会显著影响用户的实际体验（未来，计算能力会越来越强，成本参数也应相应调整）。2a12之后的前 22 位是该密文的随机盐值，最后 31 位为真正的散列值。

如果我们在数据库中存储用户密码时用的是 BCrypt 密文，那么在用户登录时，需要同步取得用户输入的密码以及数据库存储中的 BCrypt 密文，从密文中提取盐值和成本参数，与用户的密码进行一次 BCrypt 加密，最后比较两个密文是否一致。

除使用不可逆的散列算法来加密用户密码外，一些可逆的加密算法是否适合使用呢？例如，非对称加密算法 RSA，或者对称加密算法 DES、AES 等，它们看起来更加安全。是的，但正因为它们可逆，就又变得不安全了。因为当一个系统遭受"拖库"时，密钥是否还安全无法评估。事实上，我们也并不需要让用户的密码可逆。

7.3 Spring Security 的密码加密机制

Spring Security 内置了密码加密机制，只需使用一个 PasswordEncoder 接口即可。

```
public interface PasswordEncoder {
   String encode(CharSequence var1);

   boolean matches(CharSequence var1, String var2);
}
```

PasswordEncoder 接口定义了 encode 和 matches 两个方法，当用数据库存储用户密码时，加密过程用 encode 方法，matches 方法则用于判断用户登录时输入的密码是否正确。

此外，Spring Security 还内置了几种常用的 PasswordEncoder 接口，例如，NoOpPasswordEncoder、StandardPasswordEncoder 中的常规摘要算法（SHA-256 等）、BCryptPasswordEncoder 加密，以及类似 BCrypt 的慢散列加密 Pbkdf2PasswordEncoder 等，在 Spring Security 4.x 之前，默认启用 NoOpPasswordEncoder，即明文存储。

```
    public final class NoOpPasswordEncoder implements PasswordEncoder {
        private static final PasswordEncoder INSTANCE = new NoOpPasswordEncoder();

        public String encode(CharSequence rawPassword) {
            return rawPassword.toString();
        }

        public boolean matches(CharSequence rawPassword, String encodedPassword) {
            return rawPassword.toString().equals(encodedPassword);
        }
```

```
    public static PasswordEncoder getInstance() {
        return INSTANCE;
    }

    private NoOpPasswordEncoder() {
    }
}
```

当然，官方推荐我们使用 BCryptPasswordEncoder 来获得更高的安全性。

下面我们就尝试在系统中接入 BCrypt 这种密码加密方式。由于 BCrypt 加密后的密文长度超过 50，所以在我们的表结构中，如果用户密码字段过短将会导致出错，因此应首先检查并确保密码字段长度已经扩充到 60 以上。接着用一个 BCrypt 密文来实验，可以使用前面示例的 BCrypt 密文，它的原文是"blurooo"，我们可以修改表中的某个用户密码为该密文，如图 7-2 所示。

```
mysql> select * from users;
+----+----------+--------------------------------------------------------------+--------+-----------------+
| id | username | password                                                     | enable | roles           |
+----+----------+--------------------------------------------------------------+--------+-----------------+
|  1 | admin    | $2a$12$dw8zzxcArWCIucx.fOi7/O.Ky2eX/R6jTagF/6Kcly/jeQe/H0tE6 |      1 | ROLE_ADMIN,ROLE_USER |
|  2 | user     | 123                                                          |      1 | ROLE_USER       |
+----+----------+--------------------------------------------------------------+--------+-----------------+
2 rows in set (0.00 sec)
```

图 7-2

数据方面已经准备完毕，Spring Security 方面的配置其实非常简单：

```
@Bean
PasswordEncoder passwordEncoder() {
    PasswordEncoder bcryptPasswordEncoder = new BCryptPasswordEncoder(12);
    return bcryptPasswordEncoder;
}
```

在声明一个 PasswordEncoder 的 bean 之后，Spring Security 会自动应用。重启服务，发现 user 这个用户已经无法通过密码 123 来登录了。实际上需要什么密码没有人知道，但 admin 可以使用"blurooo"来登录，验证了我们的配置是有效的。

到这一步并没有完成密码加密的完整接入，因为新用户"写库"时密码还是明文，需要做一些调整。

```
// 已经定义过bean,可以直接注入
@Autowired
private PasswordEncoder passwordEncoder;

@PostMapper("/users")
public User create(User user) {
    // 加密之后重新设置
    user.setPassword(passwordEncoder.encode(user.getPasswod()));
    ...
}
```

正如 7.2 节所描述的,在一次又一次遭受破解的推力下,加密算法实际上也是一直在迭代的,尽管 BCrypt 是当前正官方所举荐的,但当未来计算机的算力已经不再是主要破解成本的时候,BCrypt 也会失去其意义,并由更强的加密算法所取代。Spring Security 作为一个强大的框架,提前考虑到算法更新换代的问题是很有必要的。

从 Spring Security 5.x 开始,官方引入了一个新的密码编码器 DelegatingPasswordEncoder(委派密码编码器),它想要解决的问题主要有三点。

◎ 确保使用当前密码存储的提议对密码进行编码。
◎ 允许同时验证现代和传统格式的密码。
◎ 允许将来升级编码。

实现上也并不困难,我们摘取其中几个比较关键的片段。

```
public DelegatingPasswordEncoder(String idForEncode, Map<String,
PasswordEncoder> idToPasswordEncoder) {
    if (idForEncode == null) {
            throw new IllegalArgumentException("idForEncode cannot be null");
    }
    if (!idToPasswordEncoder.containsKey(idForEncode)) {
            throw new IllegalArgumentException("idForEncode " +
idForEncode + "is not found in idToPasswordEncoder " + idToPasswordEncoder);
    }
    for (String id : idToPasswordEncoder.keySet()) {
            if (id == null) {
                    continue;
            }
            if (id.contains(PREFIX)) {
                    throw new IllegalArgumentException("id " + id + " cannot contain " + PREFIX);
```

```
                }
                if (id.contains(SUFFIX)) {
                    throw new IllegalArgumentException("id " + id + " cannot contain " + SUFFIX);
                }
            }
        this.idForEncode = idForEncode;
    // 新密码将使用第一个参数指明的编码器进行编码
        this.passwordEncoderForEncode = idToPasswordEncoder.get(idForEncode);
        this.idToPasswordEncoder = new HashMap<>(idToPasswordEncoder);
    }
}


// 新密码的编码存储形式为{idForEncode}encodedPassword
public String encode(CharSequence rawPassword) {
    return "{" + this.idForEncode + "}" + this.passwordEncoderForEncode.encode(rawPassword);
}


// 在密码匹配时，根据编码器 id，执行对应的编码算法
// 如果从存储的密码串中提取不到支持的编码器，则使用默认的密码编码器处理
// 默认的密码编码器实际上只是抛出该密码串不被支持的异常
public boolean matches(CharSequence rawPassword, String prefixEncodedPassword)
{
        if (rawPassword == null && prefixEncodedPassword == null) {
                return true;
        }
        String id = extractId(prefixEncodedPassword);
        PasswordEncoder delegate = this.idToPasswordEncoder.get(id);
        if (delegate == null) {
                return this.defaultPasswordEncoderForMatches
                        .matches(rawPassword, prefixEncodedPassword);
        }
        String encodedPassword = extractEncodedPassword(prefixEncodedPassword);
        return delegate.matches(rawPassword, encodedPassword);
}
```

启用新的委派密码编码方案非常简单。

```
@Bean
```

```
PasswordEncoder passwordEncoder() {
    String encodingId = "my-bcrypt";
    Map<String, PasswordEncoder> encoders = new HashMap();
    encoders.put(encodingId, new BCryptPasswordEncoder());
    return new DelegatingPasswordEncoder(encodingId, encoders);
}
```

此外，Spring Security 还提供了一个工厂类 PasswordEncoderFactories，也可以用于创建委派密码编码器。

```
@Bean
PasswordEncoder passwordEncoder() {
    PasswordEncoder passwordEncoder =
PasswordEncoderFactories.createDelegatingPasswordEncoder();
    return passwordEncoder;
}
```

通过查看源码可以得知，createDelegatingPasswordEncoder 实际是以 BCrypt 为新密码的编码算法，并兼容 Spring Security 所支持的其它几种编码类型。

```
public static PasswordEncoder createDelegatingPasswordEncoder() {
    String encodingId = "bcrypt";
    Map<String, PasswordEncoder> encoders = new HashMap();
    encoders.put(encodingId, new BCryptPasswordEncoder());
    encoders.put("ldap", new LdapShaPasswordEncoder());
    encoders.put("MD4", new Md4PasswordEncoder());
    encoders.put("MD5", new MessageDigestPasswordEncoder("MD5"));
    encoders.put("noop", NoOpPasswordEncoder.getInstance());
    encoders.put("pbkdf2", new Pbkdf2PasswordEncoder());
    encoders.put("scrypt", new SCryptPasswordEncoder());
    encoders.put("SHA-1", new MessageDigestPasswordEncoder("SHA-1"));
    encoders.put("SHA-256", new MessageDigestPasswordEncoder("SHA-256"));
    encoders.put("sha256", new StandardPasswordEncoder());
    return new DelegatingPasswordEncoder(encodingId, encoders);
}
```

通常情况下，不管是 Spring Security 4.x 使用特定密码编码器的方式还是 Spring Security 5.x 使用委派密码编码器的方式，新系统的开发都不需要考虑太多，但老系统由于存在大量旧数据，草率升级会导致老用户无法继续登录，这种情况该怎么解决呢？也并不难，这个问题我们面对的密码存储形式主要有三种。

- ◎ 明文。
- ◎ 简单 MD5 或 Spring Security 4.x 直接存储密文的形式。
- ◎ 委托密码编码器写入的携带编码器 ID 的形式。

需要特殊处理的只有前面两种，由于委派密码编码器在提取不到密码串上的编码器 ID 时，会使用默认的密码编码器算法，所以我们只需要将特殊逻辑实现为一个新的密码编码器，并设置为委派密码编码器内的默认编码器即可。

如果是明文，虽然我们可以直接以字符串对待实现兼容，但这些明文始终存在安全隐患，我们更应该考虑如何让现存的明文也得到编码，业务场景允许的情况下，可以通过跑库修改来实现，如果不允许，则可以考虑增量更新的思路，在编码器的 matches 方法中，先以明文的形式匹配，如果匹配成功，说明用户密码正确，此时可以趁机用委派密码编码器进行一次编码并更新入库，这种方案无须跑库，但无法覆盖到那些常年没有登录的用户。

如果是简单的 MD5 密码或者 Spring Security 4.x 直接存储密文的形式，同样只需要在默认的编码器中实现对应的匹配逻辑便能轻松兼容。

密码安全对系统来说从来都不是一个小的挑战，哪怕实现了安全存储，用户的密码依然会有更多的途径泄露。用户密码在提交到系统之前就被盗取的案例不胜其数，不管是制造病毒、挂马还是抓包，黑客总有办法做到。毕竟用户大多时候都不是专业的，他们往往无法准确识别陷阱，也无法安全保护自己的密码，我们能做的，只能是协助他们，以更好的安全措施来减少用户可能会犯的错误，例如采用 HTTPS 安全信道，甚至是开发银行级别的安全控件来帮助用户检测系统环境是否正常。

第 8 章
跨域与 CORS

跨域是一种浏览器同源安全策略，即浏览器单方面限制脚本的跨域访问。

8.1 认识跨域

很多人误认为资源跨域时无法请求，实际上，通常情况下请求是可以正常发起的（注意，部分浏览器存在特例），后端也正常进行了处理，只是在返回时被浏览器拦截，导致响应内容不可使用。可以论证这一点的著名案例就是 CSRF 跨站攻击。

此外，我们平常所说的跨域实际上都是在讨论浏览器行为，包括各种 WebView 容器等（其中，以 XmlHttpRequest 的使用为主）。由于 JavaScript 运行在浏览器之上，所以 Ajax 的跨域成为"痛点"。

实际上，不仅不同站点间的访问存在跨域问题，同站点间的访问可能也会遇到跨域问题，只要请求的 URL 与所在页面的 URL 首部不同即产生跨域，例如：

◎ 在 http://a.baidu.com 下访问 https://a.baidu.com 资源会形成协议跨域。
◎ 在 a.baidu.com 下访问 b.baidu.com 资源会形成主机跨域。
◎ 在 a.baidu.com:80 下访问 a.baidu.com:8080 资源会形成端口跨域。

URL 首部是指：

```
window.location.protocol + window.location.host
```

从协议部分开始到端口部分结束，只要与请求 URL 不同即被认为跨域，域名与域名对应的 IP 也不能幸免。

浏览器解决跨域问题的方法有多种，包括 JSONP、Nginx 转发和 CORS 等。其中，JSONP

和 CORS 需要后端参与。

8.2 实现跨域之 JSONP

JSONP（JSON With Padding）是一种非官方的解决方案。由于浏览器允许一些带 src 属性的标签跨域，例如，iframe、script、img 等，所以 JSONP 利用 script 标签可以实现跨域。

例如，我们有一个用于获取用户列表的 API。

```
curl -X GET \
  http://blurooo.com/users \
```

正常情况下，浏览器发出请求后服务器会给出类似下面的响应信息：

```
{
    "err_code": 0,
    "message": "",
    "data": [{
        "username": "blurooo",
        "sex": "男",
        "address": "广东深圳"
    }, {
        "username": "blurooo2"
    }]
}
```

但在跨域的情况下，浏览器的同源策略导致用户无法读取响应信息，此时前端可以使用 script 标签去加载。

```
<script src="http://blurooo.com/users"></script>
```

这样便可以成功获取响应信息了，只是得到的 JSON 数据无法直接在 JavaScript 中使用。如果后端介入，那么在返回浏览器之前应将响应信息包装成如下形式。

```
jsonp({
    "err_code": 0,
    "message": "",
    "data": [{
        "username": "blurooo",
        "sex": "男",
        "address": "广东深圳"
    }, {
```

```
        "username": "blurooo2"
    }]
})
```

对于 Java Script 而言，这就是一个普通的函数调用。

```
jsonp(...params)
```

但是 jsonp 这个函数并不存在，所以需要定义一个 jsonp 函数，以便从该函数内获取数据。

```
var jsonp = function(data){
    //输出 JSON 数据
    console.dir(data);
}
```

到这一步并不完善，因为它将导致后端无法正确处理非 JSONP 的请求，所以通常会约定一个参数 callback，带上需要包装的函数名。

```
<script src="http://blurooo.com/users?callback=jsonp"></script>
```

后端得到 callback 参数后，会使用该值包装 JSON 数据。需要注意的是，此时定义的 jsonp 函数必须在 window 对象下。

```
window.jsonp = function(data){
    console.dir(data);
}
```

如果需要挂载到别的对象下，那么与后端约定即可，这取决于后端的包装形式。通常为了更方便地使用 JSONP，前端也会做一些简单的封装。

```
var getJsonp = function(url, success){
    //声明 window 对象下的 jsonp 函数
    window.jsonp = function(data){
        //jsonp 函数被执行，将 data 转发到 success 函数
        success(data);
    }
    var src = '';
    //判断地址是否带其他参数
    if(url.IndexOf('?') != -1){
        src = url + '&callback=jsonp';
    }else{
        src = url + '?callback=jsonp';
    }
    //动态创建 script 标签
```

```
    var script = document.createElement('script');
    script.type = "text/javascript";
    script.src = src;
    document.head.appendChild(script);
}

//用法
getJsonp('http://blurooo.com/users', function(data){
    console.log('得到jsonp数据: ', JSON.stringify(data));
});
```

JSONP 的原理很简单，几乎兼容所有浏览器，实现起来也并不困难，但只支持 GET 请求跨域，局限性较大。对于部分不需要考虑兼容老旧浏览器的系统来说，CORS 的方案显得更为优雅、灵活。

8.3 实现跨域之 CORS

CORS（Cross-Origin Resource Sharing）的规范中有一组新增的 HTTP 首部字段，允许服务器声明其提供的资源允许哪些站点跨域使用。通常情况下，跨域请求即便在不被支持的情况下，服务器也会接收并进行处理，在 CORS 的规范中则避免了这个问题。浏览器首先会发起一个请求方法为 OPTIONS 的预检请求，用于确认服务器是否允许跨域，只有在得到许可后才会发出实际请求。此外，预检请求还允许服务器通知浏览器跨域携带身份凭证（如 cookie）。

CORS 新增的 HTTP 首部字段由服务器控制，下面我们来看看常用的几个首部字段：

```
Access-Control-Allow-Origin
```

允许取值为<origin>或*。<origin>指被允许的站点，使用 URL 首部匹配原则。*匹配所有站点，表示允许来自所有域的请求。但并非所有情况都简单设置即可，如果需要浏览器在发起请求时携带凭证信息，则不允许设置为*。如果设置了具体的站点信息，则响应头中的 Vary 字段还需要携带 Origin 属性，因为服务器对不同的域会返回不同的内容：

```
Access-Control-Allow-Origin: http://blurooo.com
Vary: Accept-Encoding, Origin
```

Access-Control-Allow-Methods 字段仅在预检请求的响应中指定有效，用于表明服务器允许跨域的 HTTP 方法，多个方法之间用逗号隔开。

Access-Control-Allow-Headers 字段仅在预检请求的响应中指定有效，用于表明服务器允许携带的首部字段。多个首部字段之间用逗号隔开。

Access-Control-Max-Age 字段用于指明本次预检请求的有效期，单位为秒。在有效期内，预检请求不需要再次发起。

当 Access-Control-Allow-Credentials 字段取值为 true 时，浏览器会在接下来的真实请求中携带用户凭证信息（cookie 等），服务器也可以使用 Set-Cookie 向用户浏览器写入新的 cookie。注意，使用 Access-Control-Allow- Credentials 时，Access-Control-Allow-Origin 不应该设置为*。

总体来说，CORS 是一种更安全的官方跨域解决方案，它依赖于浏览器和后端，即当需要用 CORS 来解决跨域问题时，只需要后端做出支持即可。前端在使用这些域时，基本等同于访问同源站点资源。注意，CORS 不支持 IE8 以下版本的浏览器。

在使用 CORS 时，通常有以下三种访问控制场景。

1. 简单请求

在 CORS 中，并非所有的跨域访问都会触发预检请求。例如，不携带自定义请求头信息的 GET 请求、HEAD 请求，以及 Content-Type 为 application/x-www-form-urlencoded、multipart/form-data 或 text/plain 的 POST 请求，这类请求被称为简单请求。

浏览器在发起请求时，会在请求头中自动添加一个 Origin 属性，值为当前页面的 URL 首部。当服务器返回响应时，若存在跨域访问控制属性，则浏览器会通过这些属性判断本次请求是否被允许，如果允许，则跨域成功（正常接收数据）。

```
HTTP/1.1 200 OK
...
Access-Control-Allow-Origin: http://blurooo.com
```

这种跨域请求非常简单，只需后端在返回的响应头中添加 Access-Control-Allow-Origin 字段并填入允许跨域访问的站点即可。

2. 预检请求

预检请求不同于简单请求，它会发送一个 OPTIONS 请求到目标站点，以查明该请求是否安全，防止请求对目标站点的数据造成破坏。若是请求以 GET、HEAD、POST 以外的方法发起；或者使用 POST 方法，但请求数据为 application/x-www-form-urlencoded、multipart/form-data 和 text/plain 以外的数据类型；再或者，使用了自定义请求头，则都会被当成预检请求类型处理。

3. 带凭证的请求

带凭证的请求，顾名思义，就是携带了用户 cookie 等信息的请求。

```
// XMLHttpRequest
var request = new XMLHttpRequest();
var url = 'http://blurooo.com/users';
if (request) {
   request.open('GET', url, true);
   request.withCredentials = true;
   request.onreadystatechange = function(state) {
      ...
   };
   request.send();
}

// Ajax
var settings = {
  "async": true,
  "crossDomain": true,
  "xhrFields": {
     "withCredentials": true
  },
  "url": "http://blurooo.com/users",
  "method": "GET",
}

$.ajax(settings).done(function (response) {
  console.log(response);
});
```

上面在使用 XMLHttpRequest 时，指定了 withCredentials 为 true。浏览器在实际发出请求时，将同时向服务器发送 cookie，并期待在服务器返回的响应信息中指明 Access-Control-Allow-Credentials 为 true，否则浏览器会拦截，并抛出错误。

8.4 启用 Spring Security 的 CORS 支持

事实上，Spring Security 对 CORS 提供了非常好的支持，只需在配置器中启用 CORS 支持，并编写一个 CORS 配置源即可。

```
@Override
protected void configure(HttpSecurity http) throws Exception {
```

```
    http.authorizeRequests()
        .antMatchers("/admin/api/**").hasRole("ADMIN")
        .antMatchers("/user/api/**").hasRole("USER")
        .antMatchers("/app/api/**").permitAll()
        .anyRequest().authenticated()
        .and()
    // 启用CORS支持
    .cors()
        .and()
    .formLogin().permitAll();
}

@Bean
CorsConfigurationSource corsConfigurationSource() {
    CorsConfiguration configuration = new CorsConfiguration();
    // 允许从百度站点跨域
    configuration.setAllowedOrigins(Arrays.asList("https://www.baidu.com"));
    // 允许使用GET方法和POST方法
    configuration.setAllowedMethods(Arrays.asList("GET", "POST"));
    // 允许带凭证
    configuration.setAllowCredentials(true);
    UrlBasedCorsConfigurationSource source = new
UrlBasedCorsConfigurationSource();
    // 对所有URL生效
    source.registerCorsConfiguration("/**", configuration);
    return source;
}
```

核心实现并不复杂，DefaultCorsProcessor 中的 handleInternal 方法是处理 CORS 的核心，流程非常清晰。

```
protected boolean handleInternal(ServerHttpRequest request, ServerHttpResponse
response, CorsConfiguration config, boolean preFlightRequest) throws
IOException {
    String requestOrigin = request.getHeaders().getOrigin();
    // request 中被允许的域
    String allowOrigin = this.checkOrigin(config, requestOrigin);
    HttpMethod requestMethod = this.getMethodToUse(request, preFlightRequest);
    // request 中被允许的方法
    List<HttpMethod> allowMethods = this.checkMethods(config, requestMethod);
    List<String> requestHeaders = this.getHeadersToUse(request,
preFlightRequest);
    // request 中被允许的头字段
```

```java
    List<String> allowHeaders = this.checkHeaders(config, requestHeaders);
    if(allowOrigin == null || allowMethods == null || preFlightRequest &&
allowHeaders == null) {
        this.rejectRequest(response);
        return false;
    } else {
        HttpHeaders responseHeaders = response.getHeaders();
        responseHeaders.setAccessControlAllowOrigin(allowOrigin);
        responseHeaders.add("Vary", "Origin");
        if(preFlightRequest) {
            responseHeaders.setAccessControlAllowMethods(allowMethods);
        }

        if(preFlightRequest && !allowHeaders.isEmpty()) {
            responseHeaders.setAccessControlAllowHeaders(allowHeaders);
        }

        if(!CollectionUtils.isEmpty(config.getExposedHeaders())) {
responseHeaders.setAccessControlExposeHeaders(config.getExposedHeaders());
        }

        if(Boolean.TRUE.equals(config.getAllowCredentials())) {
            responseHeaders.setAccessControlAllowCredentials(true);
        }

        if(preFlightRequest && config.getMaxAge() != null) {
responseHeaders.setAccessControlMaxAge(config.getMaxAge().longValue());
        }

        response.flush();
        return true;
    }
}
```

第 9 章
跨域请求伪造的防护

CSRF 的全称是（Cross Site Request Forgery），可译为跨域请求伪造，是一种利用用户带登录态的 cookie 进行安全操作的攻击方式。CSRF 实际上并不难防，但常常被系统开发者忽略，从而埋下巨大的安全隐患。

9.1 CSRF 的攻击过程

为了试图说明 CSRF 的攻击过程，现在思考下面这个场景。

假如有一个博客网站，为了激励用户写出高质量的博文，设定了一个文章被点赞就能奖励现金的机制，于是有了一个可用于点赞的 API，只需传入文章 id 即可：

```
http://blog.xxx.com/articles/like?id=xxx
```

在安全策略上，限定必须是本站有效登录用户才可以点赞，且每个用户对每篇文章仅可点赞一次，防止无限刷赞的情况发生。

这套机制推行起来似乎没什么问题，直我们发现有个用户的文章总是有非常多的点赞数，哪怕只是发表了一条个人状态也有非常多的点赞数，而这些点赞记录也确实都是本站的真实用户发起的。察觉到异常之后，开始对这个用户的所有行为进行排查，发现该用户几乎每篇文章都带有一张很特别的图片，这些图片的 URL 无一例外地指向了对应文章的点赞 API。由于图片是由浏览器自动加载的，所以每个查看过该文章的人都会不知不觉为其点赞。很显然，该用户利用了系统的 CSRF 漏洞实施刷赞，这是网站开发人员始料未及的。

有人可能认为这仅仅是因为点赞 API 设计不理想导致的，应当使用 POST 请求，这样就能避免上面的场景。然而，当使用 POST 请求时，确实避免了如 img、script、iframe 等标签自动发起 GET 请求的问题，但这并不能杜绝 CSRF 攻击的发生。一些恶意网站会通过表单的形式构造攻击请求：

```
<form action="http://xxx.bank.com/xxx/transfer" method="post">
    <input type="hidden" name="money" value="10000"/>
    <input type="hidden" name="to" value="hacker"/>
    <input type="submit" value="点我翻看美女图片"/>
</form>
```

假如登录过某银行站点而没有注销，其间被诱导访问了带有类似攻击的页面，那么在该页面一旦单击按钮，很可能会导致在该银行的账户资金被直接转走。甚至根本不需要单击按钮，而是直接用 JavaScript 代码自动化该过程。

CSRF 利用了系统对登录期用户的信任，使得用户执行了某些并非意愿的操作从而造成损失。如何真正地防范 CSRF 攻击，对每个有安全需求的系统而言都尤为重要。

9.2　CSRF 的防御手段

一些工具可以检测系统是否存在 CSRF 漏洞，例如，CSRFTester，有兴趣的读者可以自行了解。

在任何情况下，都应当尽可能地避免以 GET 方式提供涉及数据修改的 API。在此基础上，防御 CSRF 攻击的方式主要有以下两种。

1. HTTP Referer

HTTP Referer 是由浏览器添加的一个请求头字段，用于标识请求来源，通常用在一些统计相关的场景，浏览器端无法轻易篡改该值。

回到前面构造 POST 请求实行 CSRF 攻击的场景，其必要条件就是诱使用户跳转到第三方页面，在第三方页面构造发起的 POST 请求中，HTTP Referer 字段不是银行的 URL（少部分老版本的 IE 浏览器可以调用 API 进行伪造，但最后的执行逻辑是放在用户浏览器上的，只要用户的浏览器版本较新，便可以避免这个问题），当校验到请求来自其他站点时，可以认为是 CSRF 攻击，从而拒绝该服务。

当然，这种方式简单便捷，但并非完全可靠。除前面提到的部分浏览器可以篡改 HTTP Referer 外，如果用户在浏览器中设置了不被跟踪，那么 HTTP Referer 字段就不会自动添加，当

2. CsrfToken 认证

CSRF 是利用用户的登录态进行攻击的，而用户的登录态记录在 cookie 中。其实攻击者并不知道用户的 cookie 存放了哪些数据，于是想方设法让用户自身发起请求，这样浏览器便会自行将 cookie 传送到服务器完成身份校验。

CsrfToken 的防范思路是，添加一些并不存放于 cookie 的验证值，并在每个请求中都进行校验，便可以阻止 CSRF 攻击。

具体做法是在用户登录时，由系统发放一个 CsrfToken 值，用户携带该 CsrfToken 值与用户名、密码等参数完成登录。系统记录该会话的 CsrfToken 值，之后在用户的任何请求中，都必须带上该 CsrfToken 值，并由系统进行校验。

这种方法需要与前端配合，包括存储 CsrfToken 值，以及在任何请求中（包括表单和 Ajax）携带 CsrfToken 值。安全性相较于 HTTP Referer 提高很多，但也存在一定的弊端。例如，在现有的系统中进行改造时，前端的工作量会非常大，几乎要对所有请求进行处理。如果都是 XMLHttpRequest，则可以统一添加 CsrfToken 值；但如果存在大量的表单和 a 标签，就会变得非常烦琐。因此建议在系统开发之初考虑如何防御 CSRF 攻击。

9.3 使用 Spring Security 防御 CSRF 攻击

CSRF 攻击完全是基于浏览器进行的，如果我们的系统前端并非在浏览器中运作，就应当关闭 CSRF。

Spring Security 通过注册一个 CsrfFilter 来专门处理 CSRF 攻击。

在 Spring Security 中，CsrfToken 是一个用于描述 Token 值，以及验证时应当获取哪个请求参数或请求头字段的接口。

```
public interface CsrfToken extends Serializable {
    String getHeaderName();

    String getParameterName();

    String getToken();
}
```

CsrfTokenRepository 则定义了如何生成、保存以及加载 CsrfToken。

```java
public interface CsrfTokenRepository {
   CsrfToken generateToken(HttpServletRequest var1);

    void saveToken(CsrfToken var1, HttpServletRequest var2, HttpServletResponse var3);

   CsrfToken loadToken(HttpServletRequest var1);
}
```

在默认情况下，Spring Security 加载的是一个 HttpSessionCsrfTokenRepository。

```java
public final class HttpSessionCsrfTokenRepository implements CsrfTokenRepository {
   private static final String DEFAULT_CSRF_PARAMETER_NAME = "_csrf";
   private static final String DEFAULT_CSRF_HEADER_NAME = "X-CSRF-TOKEN";
   private static final String DEFAULT_CSRF_TOKEN_ATTR_NAME =
HttpSessionCsrfTokenRepository.class.getName().concat(".CSRF_TOKEN");
   private String parameterName = "_csrf";
   private String headerName = "X-CSRF-TOKEN";
   private String sessionAttributeName;

   public HttpSessionCsrfTokenRepository() {
       this.sessionAttributeName = DEFAULT_CSRF_TOKEN_ATTR_NAME;
   }

   public void saveToken(CsrfToken token, HttpServletRequest request, HttpServletResponse response) {
       HttpSession session;
       if(token == null) {
           session = request.getSession(false);
           if(session != null) {
               session.removeAttribute(this.sessionAttributeName);
           }
       } else {
           session = request.getSession();
           session.setAttribute(this.sessionAttributeName, token);
       }

   }

   public CsrfToken loadToken(HttpServletRequest request) {
       HttpSession session = request.getSession(false);
```

```
        return session == 
null?null:(CsrfToken)session.getAttribute(this.sessionAttributeName);
    }

    public CsrfToken generateToken(HttpServletRequest request) {
        return new DefaultCsrfToken(this.headerName, this.parameterName, 
this.createNewToken());
    }

    public void setParameterName(String parameterName) {
        Assert.hasLength(parameterName, "parameterName cannot be null or empty");
        this.parameterName = parameterName;
    }

    public void setHeaderName(String headerName) {
        Assert.hasLength(headerName, "headerName cannot be null or empty");
        this.headerName = headerName;
    }

    public void setSessionAttributeName(String sessionAttributeName) {
        Assert.hasLength(sessionAttributeName, "sessionAttributename cannot be 
null or empty");
        this.sessionAttributeName = sessionAttributeName;
    }

    private String createNewToken() {
        return UUID.randomUUID().toString();
    }
}
```

HttpSessionCsrfTokenRepository 将 CsrfToken 值存储在 HttpSession 中，并指定前端把 CsrfToken 值放在名为 "_csrf" 的请求参数或名为 "X-CSRF-TOKEN" 的请求头字段里（可以调用相应的设置方法来重新设定）。校验时，通过对比 HttpSession 内存储的 CsrfToken 值与前端携带的 CsrfToken 值是否一致，便能断定本次请求是否为 CSRF 攻击。

当使用 HttpSessionCsrfTokenRepository 时，前端必须用服务器渲染的方式注入 CsrfToken 值，例如 jsp 标签。

```
<c:url value="/login" var="loginUrl"/>
<form action="${loginUrl}" method="post">
        <p>
                <label for="username">用户名</label>
```

```
                <input type="text" id="username" name="username"/>
        </p>
        <p>
                <label for="password">密码</label>
                <input type="password" id="password" name="password"/>
        </p>
        <input type="hidden"
                name="${_csrf.parameterName}"
                value="${_csrf.token}"/>
        <button type="submit" class="btn">登录</button>
</form>
```

这种方式在某些单页应用中局限性较大，灵活性不足。

Spring Security 还提供了另一种方式，即 CookieCsrfTokenRepository。

```
public final class CookieCsrfTokenRepository implements CsrfTokenRepository {
    static final String DEFAULT_CSRF_Cookie_NAME = "XSRF-TOKEN";
    static final String DEFAULT_CSRF_PARAMETER_NAME = "_csrf";
    static final String DEFAULT_CSRF_HEADER_NAME = "X-XSRF-TOKEN";
    private String parameterName = "_csrf";
    private String headerName = "X-XSRF-TOKEN";
    private String cookieName = "XSRF-TOKEN";
    private final Method setHttpOnlyMethod;
    private boolean cookieHttpOnly;
    private String cookiePath;

    ...

    public void saveToken(CsrfToken token, HttpServletRequest request,
HttpServletResponse response) {
        String tokenValue = token == null?"":token.getToken();
        Cookie cookie = new Cookie(this.cookieName, tokenValue);
        cookie.setSecure(request.isSecure());
        if(this.cookiePath != null && !this.cookiePath.isEmpty()) {
           cookie.setPath(this.cookiePath);
        } else {
           cookie.setPath(this.getRequestContext(request));
        }

        if(token == null) {
           cookie.setMaxAge(0);
        } else {
           cookie.setMaxAge(-1);
```

```
        }

        if(this.cookieHttpOnly && this.setHttpOnlyMethod != null) {
            ReflectionUtils.invokeMethod(this.setHttpOnlyMethod, cookie, new 
Object[]{Boolean.TRUE});
        }

        response.addCookie(cookie);
    }
    ...
}
```

CookieCsrfTokenRepository 是一种更加灵活可行的方案，它将 CsrfToken 值存储在用户的 cookie 内。首先，减少了服务器 HttpSession 存储的内存消耗；其次，当用 cookie 存储 CsrfToken 值时，前端可以用 JavaScript 读取（需要设置该 cookie 的 httpOnly 属性为 false），而不需要服务器注入参数，在使用方式上更加灵活。

有的读者可能会有疑惑，存储在 cookie 上，不就又可以被 CSRF 利用了吗？事实上并不可以。cookie 只有在同域的情况下才能被读取，所以杜绝了第三方站点跨域获取 CsrfToken 值的可能。CSRF 攻击本身是不知道 cookie 内容的，只是利用了当请求自动携带 cookie 时可以通过身份验证的漏洞。但服务器对 CsrfToken 值的校验并非取自 cookie，而是需要前端手动将 CsrfToken 值作为参数携带在请求里，所以 cookie 内的 CsrfToken 值并没有被校验的作用，仅仅作为一个存储容器使用。

修改 Spring Security 的 csrfTokenRepository。

```
@Override
protected void configure(HttpSecurity http) throws Exception {
    http.csrf()
        .csrfTokenRepository(CookieCsrfTokenRepository.withHttpOnlyFalse());
}
```

了解了 CsrfToken 和 csrfTokenRepository 之后，再来看看 csrfFilter。

```
protected void doFilterInternal(HttpServletRequest request, 
HttpServletResponse response, FilterChain filterChain) throws ServletException, 
IOException {
    request.setAttribute(HttpServletResponse.class.getName(), response);
    CsrfToken csrfToken = this.tokenRepository.loadToken(request);
    boolean missingToken = csrfToken == null;
```

```java
    if(missingToken) {
        csrfToken = this.tokenRepository.generateToken(request);
        this.tokenRepository.saveToken(csrfToken, request, response);
    }

    request.setAttribute(CsrfToken.class.getName(), csrfToken);
    request.setAttribute(csrfToken.getParameterName(), csrfToken);
    if(!this.requireCsrfProtectionMatcher.matches(request)) {
        filterChain.doFilter(request, response);
    } else {
        String actualToken = request.getHeader(csrfToken.getHeaderName());
        if(actualToken == null) {
            actualToken = request.getParameter(csrfToken.getParameterName());
        }

        if(!csrfToken.getToken().equals(actualToken)) {
            if(this.logger.isDebugEnabled()) {
                this.logger.debug("Invalid CSRF token found for " +
                            UrlUtils.buildFullRequestUrl(request));
            }

            if(missingToken) {
                this.accessDeniedHandler.handle(request, response, new
                    MissingCsrfTokenException(actualToken));
            } else {
                this.accessDeniedHandler.handle(request, response, new
                    InvalidCsrfTokenException(csrfToken, actualToken));
            }

        } else {
            filterChain.doFilter(request, response);
        }
    }
}
```

csrfFilter 的处理流程很清晰，当一个请求到达时，首先会调用 csrfTokenRepository 的 loadToken 方法加载该会话的 CsrfToken 值。如果加载不到，则证明请求是首次发起的，应该生成并保存一个新的 CsrfToken 值。如果可以加载到 CsrfToken 值，那么先排除部分不需要验证 CSRF 攻击的请求方法（默认忽略了 GET、HEAD、TRACE 和 OPTIONS）。

```java
private static final class DefaultRequiresCsrfMatcher implements RequestMatcher
{
    private final HashSet<String> allowedMethods;
```

```
    private DefaultRequiresCsrfMatcher() {
       this.allowedMethods = new HashSet(Arrays.asList(new String[]{"GET",
           "HEAD", "TRACE", "OPTIONS"}));
    }

    public boolean matches(HttpServletRequest request) {
       return !this.allowedMethods.contains(request.getMethod());
    }
}
```

当请求确认需要验证时,获取其携带的 CsrfToken 值,并与前面加载到的 CsrfToken 值进行比较即可。

2016 年,Spring Security 社区有人指出,csrfFilter 总是在创建会话时,触发生成并保存一个 CsrfToken 值,即便该会话实际上用不到这个 CsrfToken 值。例如,当只是使用一些公开的 GET 类型 API 时,既不需要身份验证,也不需要 CSRF 攻击验证,那么此时保存的 Csrftoken 值就是浪费空间资源。

于是 Spring Security 新增了一个 LazyCsrfTokenRepository,用来延时保存 CsrfToken 值(允许创建,但只有真正使用时才会被保存)。

```
public final class LazyCsrfTokenRepository implements CsrfTokenRepository {
    private static final String HTTP_RESPONSE_ATTR =
        HttpServletResponse.class.getName();
    private final CsrfTokenRepository delegate;

    public LazyCsrfTokenRepository(CsrfTokenRepository delegate) {
       Assert.notNull(delegate, "delegate cannot be null");
       this.delegate = delegate;
    }

    public CsrfToken generateToken(HttpServletRequest request) {
       return this.wrap(request, this.delegate.generateToken(request));
    }

    public void saveToken(CsrfToken token, HttpServletRequest request,
      HttpServletResponse response) {
       if(token == null) {
          this.delegate.saveToken(token, request, response);
       }
```

```java
    }

    public CsrfToken loadToken(HttpServletRequest request) {
        return this.delegate.loadToken(request);
    }

    private CsrfToken wrap(HttpServletRequest request, CsrfToken token) {
        HttpServletResponse response = this.getResponse(request);
        return new LazyCsrfTokenRepository.SaveOnAccessCsrfToken(this.delegate,
            request, response, token);
    }

    private HttpServletResponse getResponse(HttpServletRequest request) {
        HttpServletResponse response =
            (HttpServletResponse)request.getAttribute(HTTP_RESPONSE_ATTR);
        if(response == null) {
            throw new IllegalArgumentException("The HttpServletRequest attribute
                                        must contain an HttpServletResponse
                                        for the attribute " +
                                        HTTP_RESPONSE_ATTR);
        } else {
            return response;
        }
    }

    private static final class SaveOnAccessCsrfToken implements CsrfToken {
        private transient CsrfTokenRepository tokenRepository;
        private transient HttpServletRequest request;
        private transient HttpServletResponse response;
        private final CsrfToken delegate;

        SaveOnAccessCsrfToken(CsrfTokenRepository tokenRepository,
                        HttpServletRequest request, HttpServletResponse
                        response, CsrfToken delegate) {
            this.tokenRepository = tokenRepository;
            this.request = request;
            this.response = response;
            this.delegate = delegate;
        }

        public String getHeaderName() {
            return this.delegate.getHeaderName();
        }
```

```java
public String getParameterName() {
    return this.delegate.getParameterName();
}

public String getToken() {
    this.saveTokenIfNecessary();
    return this.delegate.getToken();
}

public String toString() {
    return "SaveOnAccessCsrfToken [delegate=" + this.delegate + "]";
}

public int hashCode() {
    int prime = true;
    int result = 1;
    int result = 31 * result + (this.delegate ==
                                null?0:this.delegate.hashCode());
    return result;
}

public boolean equals(Object obj) {
    if(this == obj) {
        return true;
    } else if(obj == null) {
        return false;
    } else if(this.getClass() != obj.getClass()) {
        return false;
    } else {
        LazyCsrfTokenRepository.SaveOnAccessCsrfToken other =
            (LazyCsrfTokenRepository.SaveOnAccessCsrfToken)obj;
        if(this.delegate == null) {
            if(other.delegate != null) {
                return false;
            }
        } else if(!this.delegate.equals(other.delegate)) {
            return false;
        }

        return true;
    }
}

private void saveTokenIfNecessary() {
```

```
            if(this.tokenRepository != null) {
                synchronized(this) {
                    if(this.tokenRepository != null) {
                        this.tokenRepository.saveToken(this.delegate,
                            this.request, this.response);
                        this.tokenRepository = null;
                        this.request = null;
                        this.response = null;
                    }
                }
            }
        }
    }
}
```

可以看到，LazyCsrfTokenRepository 并非独立使用一个 csrfTokenRepository，而是专门用于包裹其他 csrfTokenRepository。LazyCsrfTokenRepository 先是覆盖了原 csrfTokenRepository 的 saveToken 方法，使得 csrfFilter 中的 saveToken 方法失去实际的保存效果；接着又修改了 generateToken，使得 CsrfToken 在首次调用 getToken 时，才真正调用 saveToken 方法对 CsrfToken 进行保存。此特性发布在 Spring Security 4.1.0.RELEASE 版本中，在该版本之后，我们看到的 csrfConfigurer 已经默认使用 LazyCsrfTokenRepository 来包裹 HttpSessionCsrfTokenRepository。

```
public final class CsrfConfigurer<H extends HttpSecurityBuilder<H>> extends
AbstractHttpConfigurer<CsrfConfigurer<H>, H> {
    private CsrfTokenRepository csrfTokenRepository = new
        LazyCsrfTokenRepository(new HttpSessionCsrfTokenRepository());

    ...

}
```

第 10 章
单点登录与CAS

单点登录（Single Sign On，SSO）是指在多个应用系统中，只需登录一次，即可同时以登录态共享企业所有相关又彼此独立的系统的功能。对于旗下拥有众多系统的企业来说，单点登录不仅降低了用户的登录成本，统一了不同系统间的账号体系，还减少了各个系统在用户设计上付出的精力。

10.1 单点登录

作为互联网用户，我们一直在享受单点登录这项技术带来的便利。如谷歌公司的主域名是提供搜索服务的，旗下有邮箱和地图等知名应用，我们只需在任意一个应用内登录一次，即可使用所有这些服务，非常方便。除谷歌外，国内也有类似的案例，如阿里巴巴，我们同样只需在阿里巴巴旗下的淘宝、天猫、1688 等任意系统中登录一次，即可将登录态同步到其他系统中，甚至这些系统的域名完全不同，如 taobao.com、tmall.com 和 1688.com。

这是如何实现的呢？在此之前，我们先来理解单个应用的登录方案。

前面简单介绍过会话的机制，服务器在接收请求后，会为每个新用户生成一个会话 ID，该会话 ID 不仅绑定了用户信息，还会被设置到用户的浏览器中。由于浏览器每次发起请求时都会自动携带 cookie，所以服务器可以通过 cookie 获取会话 ID，从而找到请求对应的用户，大致交互如图 10-1 所示。

设置 cookie 是在 HTTP 中进行的，只需在响应体中添加首部信息：Set-Cookie，浏览器便会自动解析并储存 cookie。

```
Set-Cookie: <cookie-name>=<cookie-value>; Domain=<domain-value>;
Path=<path-value>; Secure; HttpOnly
```

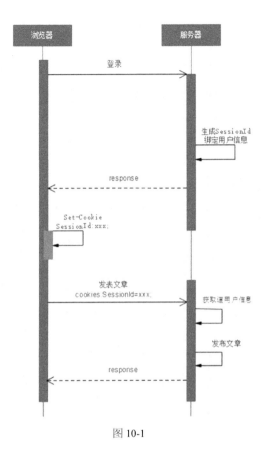

图 10-1

Web 框架一般也会提供 cookie 的设置方法，并且不需要开发者了解其交互细节。例如，使用 Servlet 设置 cookie。

```
Cookie cookie = new Cookie("spring.security", "blurooo");
response.addCookie(cookie);
```

我们可以在 Chrome 浏览器中查看该 cookie 的信息，如图 10-2 所示。

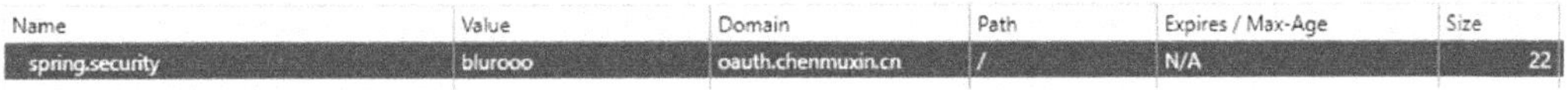

Name	Value	Domain	Path	Expires / Max-Age	Size
spring.security	blurooo	oauth.chenmuxin.cn	/	N/A	22

图 10-2

在互联网中，几乎每个域名都有自己的 cookie，如果每一个请求都携带浏览器存储的所有 cookie，那么使用的互联网服务越多，积累的 cookie 就越多。而 cookie 越多，则每次请求所要传输的流量就越大，获得的服务响应时间就越长，最终浏览器和互联网都将不堪重负。因此，

cookie 的 Domain 设置至关重要。

Domain 实际上圈定了 cookie 的作用范围。例如，我们在访问 oauth.chenmuxin.cn 时会获得一个 cookie，该 cookie 的 Domain 为 oauth.chenmuxin.cn。只有在访问 oauth.chenmuxin.cn 或 xx.oauth.chenmuxin.cn 时，浏览器才会将该 cookie 传输到服务器。如果访问的是 chenmuxin.cn 或 sso.chenmuxin.cn，那么该 cookie 不会生效。相反，在设置 cookie 时，允许将 Domain 指定为当前域名或当前域名的所有上级域名。例如，oauth.chenmuxin.cn 仅允许将 cookie 的 Domain 设置为 oauth.chenmuxin.cn 或其上级域名 chenmuxin.cn，其他 Domain 将被浏览器拒绝。

如果用户在 mail.google.com 中完成了登录，那么表示会话 ID 的 cookie 的 Domain 一般会被设置为 mail.google.com。也就是说，会话 ID 仅在访问 mail.google.com 或其子域名时才会生效，无法共享给 google.com 或 maps.google.com。

如果将 Domain 设置为 google.com，那么是不是所有 google.com 下的应用都可以共享了呢？的确如此，但还有些问题需要考虑。假如我们把会话 ID 设置到顶级域名下，使该顶级域名的所有子域名都可以共享，但却并非所有系统都能识别。因为会话 ID 是由该域名下的某个服务生成的，会话的数据通常只存储在该服务中，为了解决这个问题，还需要引入共享会话的方案。共享会话是实现会话集群管理的一种比较推荐的方案，原理也很简单，只需将这些系统的会话数据抽离并存储在独立的数据容器中即可。

将 cookie 的域设置为顶级域名解决了 cookie 共享问题，但这对不同域名下的系统是行不通的。出于安全考虑，cookie 无法在服务器实现跨域设置，即，在 taobao.com 下无法直接将 cookie 设置到 tmall.com。即便可以设置，也会存在一个问题：每次修改 cookie 时都需要将 cookie 再次同步到所有其他系统中，一旦有新增或删减，那么每个系统都必须修改发版。

阿里巴巴旗下的系统是如何解决这个问题的呢？实际上，这些系统有一个统一的登录服务：login.taobao.com，用户首次登录淘宝时会请求得到一个类似于会话 ID 的 cookie 字段 _tb_token_，其 Domian 被设置为 taobao.com。当天猫某个被访问的 target 页面需要共享淘宝的登录态时，可以先跳转到 https://login.taobao.com/jump?target=https://tmall.com/xxx?tbpm=1，由于 login.taobao.com 属于 taobao.com 下的二级域名，所以 _tb_token_ 将生效并被传送到服务器，服务器接收请求之后，会解析 target 对应的系统，将淘宝所有与共享登录态相关的 cookie 转换为查询参数，拼接在对应系统的 cookie 设置页面上并执行重定向。cookie 设置页面一般为 https://pass.xxx.com/add。例如，在这个案例中，会重定向到 https://pass.tmall.com/add?tb_token=xxx&...&target=https://tmall.com/xxx?tbpm=1。这个页面将从查询参数中解析 cookie，并设置到 tmall.com 这个域下，最后再次通过重定向回到 target 页面。整个交互流程如图 10-3 所示。

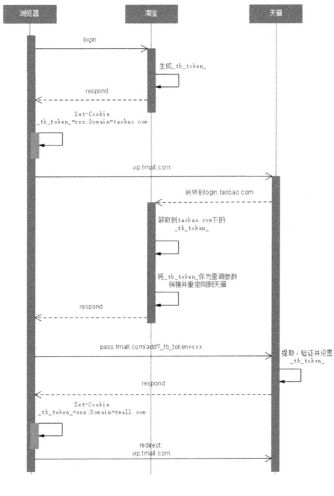

图 10-3

10.2　认识 CAS

如果企业旗下的所有系统都使用同样的顶级域名，那么实现单点登录就会变得非常简单，只需将 cookie 的域设置为顶级域名，在服务器使用共享会话的方案即可。但很多时候并没有这么理想的状态，因此实现单点登录的成本相对较高。

实际上，开源社区提供了一套非常好的系统：CAS（Central Authentication Service，中央验证服务），利用 CAS 实现单点登录将大大降低开发及维护成本。

CAS 由 CAS Server 和 CAS Client 两部分组成。CAS Server 是一个单点的验证服务，CAS Client 是共享 CAS Server 登录态的客户端。例如，阿里巴巴旗下的淘宝、天猫，在 CAS 结构中都属于 CAS Client。此外，CAS 还有三个重要的术语：Ticket Granting Ticket（TGT）、Ticket Granted Cookie（TGC）和 Service Ticket（ST）。

- ◎ TGT 是用户登录后生成的票根，包含用户的认证身份、有效期等，存储于 CAS Server 中，类似于我们常见的服务器会话。
- ◎ TGC 是存储在 cookie 中的一段数据，类似于会话 ID，用户与 CAS Server 进行交互时，帮助用户找到对应的 TGT。
- ◎ ST 是 CAS Server 使用 TGT 签发的一张一次性票据，CAS Client 使用 ST 与 CAS Server 进行交互，以获取用户的验证状态。

AS 单点登录的完整步骤如下：

（1）用户通过浏览器访问 CAS Client 的某个页面，例如 http://cas.client.com/me。

（2）当 CAS Client 判断用户需要进行身份认证时，携带 service 返回 302 状态码，指示浏览器重定向到 CAS Server，例如 http://cas.server.com/?service=http://cas.client.com/me。service 指用户原访问页面。

（3）浏览器携带 service 重定向到 CAS Server。

（4）CAS Server 获取并校验用户 cookie 中携带的 TGC，如果成功，则身份认证完成；否则将用户重定向到 CAS Server 提供的登录页，例如 http://cas.server.com/login?service=http://cas.client.com/me，由用户输入用户名和密码，完成身份认证。

（5）如果用户已经登录过系统，那么 CAS Server 可以获取用户的 TGC，并根据 TGC 找到 TGT。如果是首次登录，则 CAS Server 会首先生成 TGT。每次验证时，CAS Server 会根据 TGT 签发一个 ST，把 ST 拼接在 service 中，同时将相应的 TGC 设置到用户的 cookie 中（域为 CAS Server），并返回 302 状态码，指示浏览器重定向到 service，例如 http://cas.client.com/me?ticket=xxx。

（6）浏览器存储 TGC，并携带 ST 重定向到 service。

（7）CAS Client 取得 ST（即请求参数中的 ticket）后，会向 CAS Server 请求验证该 ST 的有效性。

（8）若 CAS Server 验证该 ST 是有效的，就告知 CAS Client 该用户有效，并返回该用户的信息。

CAS Client 在获取用户信息时，可以使用 session 的形式管理用户会话。后续的交互请求不再需要重定向到 CAS Server，CAS Client 直接返回用户请求的资源即可。整个流程如图 10-4 所示。

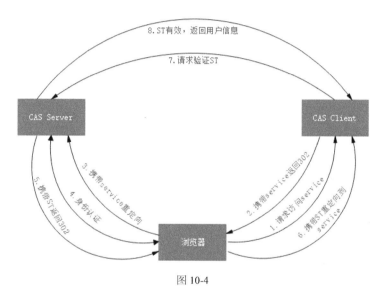

图 10-4

10.3 搭建 CAS Server

为了验证 CAS 单点登录，我们需要在本地开发环境中搭建一套用于测试的 CAS Server，笔者基于 Maven 构建 CAS Server5.3 版本。

1. 环境依赖

- JDK1.8+，需要配置好环境变量，在搭建过程中将使用 Java 提供的 keytool 证书制作工具。
- Maven，需要配置好环境变量，在 Windows 的 CMD 或 macOS X 的 Terminal 中输入"mvn –v"进行验证。

2. 导入 CAS Server 项目

CAS Server 的搭建基于 GitHub 项目：https://github.com/apereo/cas-overlay-template。将这个项目导入 IDEA，待下载完所有依赖的 maven 插件以及最关键的 cas-overlays 后（可能会有部分插件无法导入，影响不大，忽略即可），在项目目录结构上会出现 overlays 目录，如图 10-5 所示。

图 10-5

3. 制作本地密钥库

CAS Server 默认使用 HTTPS 进行访问，并要求我们提供一个密钥库。

可以使用 Java 自带的密钥和证书管理工具 keytool 制作本地密钥库，它位于 {JAVA_HOMR}/bin 目录下。在配置环境变量时，在控制台直接输入 keytool 命令就可以获取帮助，如图 10-6 所示。

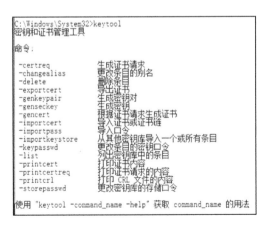

图 10-6

子命令 genkeypair 可生成密钥对，使用方式如图 10-7 所示。

```
C:\Windows\System32>keytool -genkeypair -help
keytool -genkeypair [OPTION]...

生成密钥对

选项:

 -alias <alias>                  要处理的条目的别名
 -keyalg <keyalg>                密钥算法名称
 -keysize <keysize>              密钥位大小
 -sigalg <sigalg>                签名算法名称
 -destalias <destalias>          目标别名
 -dname <dname>                  唯一判别名
 -startdate <startdate>          证书有效期开始日期/时间
 -ext <value>                    X.509 扩展
 -validity <valDays>             有效天数
 -keypass <arg>                  密钥口令
 -keystore <keystore>            密钥库名称
 -storepass <arg>                密钥库口令
 -storetype <storetype>          密钥库类型
 -providername <providername>    提供方名称
 -providerclass <providerclass>  提供方类名
 -providerarg <arg>              提供方参数
 -providerpath <pathlist>        提供方类路径
 -v                              详细输出
 -protected                      通过受保护的机制的口令

使用 "keytool -help" 获取所有可用命令
```

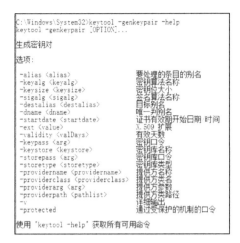

图 10-7

-alias 是要处理的条目的别名，可以根据个人喜好设定。

-keyalg 用来指定证书密钥算法名称，支持 DSA 和 RSA。

-keystore 用来指定密钥库名称，若不存在，则需要新建。

制作本地密钥库。

```
keytool -genkey -alias casserver -keyalg RSA -keystore D:\keystore
```

可以根据实际情况修改 keystore 的存储位置，完整的交互过程如图 10-8 所示。

```
C:\Windows\System32>keytool -genkey -alias casserver -keyalg RSA -keystore D:\keystore
输入密钥库口令:
再次输入新口令:
您的名字与姓氏是什么?
  [Unknown]:  cas.chenmuxin.cn
您的组织单位名称是什么?
  [Unknown]:
您的组织名称是什么?
  [Unknown]:
您所在的城市或区域名称是什么?
  [Unknown]:
您所在的省/市/自治区名称是什么?
  [Unknown]:
该单位的双字母国家/地区代码是什么?
  [Unknown]:
CN=cas.chenmuxin.cn, OU=Unknown, O=Unknown, L=Unknown, ST=Unknown, C=Unknown是否正确?
  [否]:  y

输入 <casserver> 的密钥口令
        (如果和密钥库口令相同, 按回车):

Warning:
JKS 密钥库使用专用格式。建议使用 "keytool -importkeystore -srckeystore D:\keystore -destkeystore D:\keystore -deststoret
ype pkcs12" 迁移到行业标准格式 PKCS12。
```

图 10-8

需要注意的是，在交互过程中会涉及两个口令：密钥库口令和密钥口令。如果在生成密钥时，keystore 指定的密钥库已经存在，那么新生成的密钥就会添加到该密钥库，否则将生成一个新的密钥库用于存放新生成的密钥。所以，当指定的密钥库已经存在时，必须填写指定密钥库的口令；如果密钥库不存在，则可以设定一个新口令。

另外，在"您的名字与姓氏是什么？"一项，应当填写 CAS Server 的域名，否则在后续的单点登录过程中会遇到问题。如果只是用于本地开发测试，则域名可以随意填写，并通过配置 hosts 的方式使其生效。其他参数可以根据个人情况填写，或者直接跳过。制作成功之后就可以在对应的路径下找到密钥库了。

因为后续需要把 Spring Security 接入 CAS Server，所以作为客户端，在与 HTTPS 服务进行通信时，不得不考虑 SSL 证书的问题，以免抛出下面这种错误。

```
javax.net.ssl.SSLHandshakeException:
sun.security.validator.ValidatorException: PKIX path building failed:
sun.security.provider.certpath.SunCertPathBuilderException: unable to find
valid certification path to requested target
```

其实，CAS Server 的证书问题并非客户端需要考虑的，在真实的接入场景中，CAS Server 是支持 HTTPS 的，但因为自制证书会影响开发测试，所以必须解决这个问题。实际上，我们只需将证书导入 JDK 中即可，更多信息请参考官方的说明文档。

JDK 自带了一个证书库，在 JDK8 及以前的版本中，该证书库位于%JAVAHOME%/jre/lib/security/cacerts 中。从 JDK9 开始，取消了 jre 目录，因此证书库位于%JAVAHOME%/lib/security/cacerts 中，我们需要从前面制作的密钥库中导出证书，并将其导入 cacerts 中。

使用 export 子命令导出证书。

```
keytool -export -trustcacerts -alias casserver -file D:\cas.cer -keystore
D:\keystore
```

执行效果是：从 D:\keystore 这个密钥库中，导出别名为 casserver 的证书到 D:\cas.cer 文件中。密钥库口令为 keystore 生成时自定义的口令。

使用 import 子命令导入证书。

```
keytool -import -trustcacerts -alias casserver -file D:\cas.cer -keystore
"C:\Program Files (x86)\Java\jdk1.8.0_181\jre\lib\security\cacerts"
```

执行效果是：以 casserver 作为别名，把 D:\cas.cer 这个证书文件导入 C:\Program Files (x86)\Java\jdk1.8.0_181\jre\lib\security\cacerts 证书库中（具体路径请参照本地开发环境中配置好的 JDK 路径）。密钥库口令为 cacerts 的默认口令：changeit。

4. 覆盖 CAS Server 原配置

CAS Server 是基于 overlays 的方式构建的。overlays 是一种对抗重复代码和资源的策略，它允许我们在多个 Web 工程中共享通用资源。如果仅仅搭建一个本地测试的 CAS Server，那么只需覆盖一些配置即可，如图 10-9 所示。

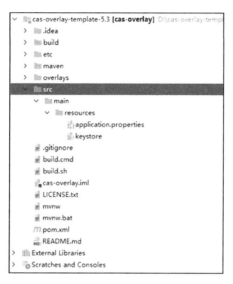

图 10-9

新建目录 src/main/resources，将 overlays/{cas-server}/WEB-INF/classes/application.properties 复制到此目录下，后续将用这个新的配置文件覆盖 CAS Server 的配置。

在 CAS Server 配置中，关于 SSL 证书的三个主要配置如下。

```
server.ssl.key-store=file:/etc/cas/thekeystore
server.ssl.key-store-password=changeit
server.ssl.key-password=changeit
```

key-store 指定密钥库的位置，key-store-password 指定密钥库的口令，key-password 指定密钥的口令。想要让 CAS Server 启动成功，则必须提供一个正确的密钥库位置。由于上个步骤生成的密钥库位于 D:\keystore 下，所以可以将 key-store 修改为下面的形式。

```
server.ssl.key-store=file:D:\\keystore
```

也可以把 keystore 复制到 resources 目录下，把 key-store 修改为下面的形式。

```
server.ssl.key-store=classpath:keystore
```

把密钥库口令和密钥口令对应修改为正确的即可。

5. 启动 CAS Server

既可以将 CAS Server 打包为 war 包的形式运行，也可以使用 Spring Boot 的方式运行，出于简单考虑，推荐采用后者。

官方分别为 Windows 和 maxOS X 系统提供了对应的运行脚本，但兼容性较差，在各类错综复杂的开发环境中难以保证成功，因此笔者对流程进行了简化，只要确保环境配置没有问题，则运行失败的情况就会大大减少。打开 IDEA 左下角的 Terminal 窗口，输入下面的命令：

```
mvn spring-boot:run
```

在启动过程中可能会报错，但不影响正常使用，只要在控制台中显示 ready 图案即可（如果是在 macOS X 中开发，则还需要将项目 etc 目录下的内容复制到文件系统的 /etc 目录下）。

6. 访问 CAS Server

前面在制作密钥库时使用了 cas.chenmuxin.cn 作为域名，为了让它生效，需要将该域名配置到本地 hosts 文件中，并指向 127.0.0.1。完成之后就可以通过 https://cas.chenmuxin.cn/cas 访问 CAS Server 了，效果如图 10-10 所示。

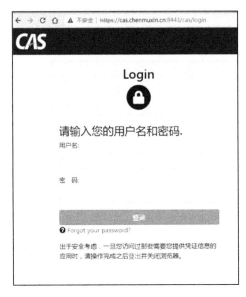

图 10-10

默认情况下使用用户名 casuser 和密码 Mellon 即可登录，如果需要修改，则可以在 application.properties 文件的最后一行中变更 cas.authn.accept.users 的值，重启后即可生效。

```
cas.authn.accept.users=casuser::Mellon
```

7. Services 配置

如果 CAS Client 想要接入某个 CAS Server，则必须定义服务记录，否则客户端在访问时将提示"未验证授权的服务"。CAS Server 本身支持多种注册服务方式，每个服务对应一个允许接入的客户端。

在测试阶段，只使用 JSON 声明的方式注册服务，每个 JSON 文件对应一个服务，建议用下面的规则创建 JSON 文件名。

```
JSON fileName = serviceName + "-" + serviceNumericId + ".json"
```

每个 JSON 文件的内容大致如下。

```
{
  "@class": "org.apereo.cas.services.RegexRegisteredService",
  "serviceId": "^https://client.cas.chenmuxin.cn",
  "name": "cas client",
  "id": 10000,
  "description": "cas client based on spring security",
  "evaluationOrder": 1
}
```

其中，@class 必须指定为 org.apereo.cas.services.RegexgisteredService 的实现类，serviceId 则通过正则表达式的方式匹配来自 CAS Client 的 URL，每个 id 都是全局唯一编号，evaluationOrder 指明了服务的执行顺序。

把该 JSON 文件命名为 cas_client-10000.json，并放置到/src/main/resources/services 下，但暂时不会生效，还需要为 CAS Server 添加一些相关配置。

```
# JSON 资源加载路径
cas.serviceRegistry.json.location=classpath:/services
# 是否开启 JSON 文件识别，默认为 false
cas.serviceRegistry.initFromJson=true
# 是否自动扫描服务配置，默认开启（在 CAS Server 启动时通过控制台定时输出的信息
# 也能观察到这一点）
# cas.serviceRegistry.watcherEnabled=true
# 默认每 120s 扫描一次服务配置
```

```
# cas.serviceRegistry.repeatInterval=120000
# 默认在 CAS Server 启动后延迟 15s 开启自动扫描
# cas.serviceRegistry.startDelay=15000
```

由于 JSON 资源加载路径被配置到了类路径下的 services 文件夹，所以在配置生效后，系统可以自动扫描该目录下的服务配置信息。重启服务，控制台将打印类似下面的信息，表明 CAS Server 已经读取并加载了服务。

```
2019-04-06 01:12:17,018 INFO [org.apereo.cas.services.AbstractServicesManager]
- <Loaded [1] service(s) from [InMemoryServiceRegistry].>
```

10.4　用 Spring Security 实现 CAS Client

在实现 CAS Client 过程中，先梳理当前准备好的两个域名信息：

◎　客户端域名为 client.cas.chenmuxin.cn，配置 hosts，指向 127.0.0.1。
◎　服务器域名为 cas.chenmuxin.cn，配置 hosts，指向 127.0.0.1。

当然，在具体操作时可以依据个人喜好修改这些域名，但从生成证书部分开始，域名就应该都确定好，避免因为域名不匹配出现问题。

1. 引入依赖包

Spring Security 提供了一个独立的依赖包来实现 CAS Client，我们只需在 pom.xml 文件中声明引入即可。

```xml
<dependency>
    <groupId>org.springframework.security</groupId>
    <artifactId>spring-security-cas</artifactId>
</dependency>
```

2. CAS Client 的相关配置信息

在与 CAS Server 进行交互时，需要配置下面这些信息。

```
# CAS Server 主路径
cas.server.prefix=https://cas.chenmuxin.cn:8443/cas
# CAS Server 登录 URL
cas.server.login=${cas.server.prefix}/login
# CAS Server 注销 URL
cas.server.logout=${cas.server.prefix}/logout
```

```
cas.client.prefix=http://client.cas.chenmuxin.cn:8080
# CAS Client 登录 URL
cas.client.login=${cas.client.prefix}/login/cas
# CAS Client 注销 URL 的相对路径形式
cas.client.logout.relative=/logout/cas
# CAS Client 注销 URL 的绝对路径形式
cas.client.logout=${cas.client.prefix}${cas.client.logout.relative}

# CAS Server 的登录用户名
cas.user.inmemory=casuser
```

CAS Client 的相关配置类如下表示。

```
@Configuration
public class CasSecurityApplication {

    @Value("${cas.server.prefix}")
    private String casServerPrefix;

    @Value("${cas.server.login}")
    private String casServerLogin;

    @Value("${cas.server.logout}")
    private String casServerLogout;

    @Value("${cas.client.login}")
    private String casClientLogin;

    @Value("${cas.client.logout}")
    private String casClientLogout;

    @Value("${cas.client.logout.relative}")
    private String casClientLogoutRelative;

    @Value("${cas.user.inmemory}")
    private String casUserInMemory;

    /**
     *
     * 配置 CAS Client 的属性
     *
     * @return
     */
```

```java
@Bean
public ServiceProperties serviceProperties() {
    ServiceProperties serviceProperties = new ServiceProperties();
    // 与 CasAuthenticationFilter 监视的 URL 一致
    serviceProperties.setService(casClientLogin);
    // 是否关闭单点登录，默认为 false，可以不设置
    serviceProperties.setSendRenew(false);
    return serviceProperties;
}

/**
 *
 * CAS 验证入口，提供用户浏览器重定向的地址
 *
 * @param sp
 * @return
 */
@Bean
@Primary
public AuthenticationEntryPoint authenticationEntryPoint(ServiceProperties sp) {
    CasAuthenticationEntryPoint entryPoint = new CasAuthenticationEntryPoint();
    // CAS Server 认证的登录地址
    entryPoint.setLoginUrl(casServerLogin);
    entryPoint.setServiceProperties(sp);
    return entryPoint;
}

/**
 *
 * ticket 校验，需要提供 CAS Server 校验 ticket 的地址
 *
 * @return
 */
@Bean
public TicketValidator ticketValidator() {
    // 默认使用 Cas20ProxyTicketValidator，验证入口是
    // ${casServerPrefix}/proxyValidate
    return new Cas20ProxyTicketValidator(casServerPrefix);
}

/**
```

```java
     *
     * 使用内存上的用户并分配权限
     *
     * @return
     */
    @Bean
    public UserDetailsService userDetailsService() {
        InMemoryUserDetailsManager manager = new InMemoryUserDetailsManager();
manager.createUser(User.withUsername(casUserInMemory).password("").roles("USER").build());
        return manager;
    }

    /**
     * CAS 验证处理逻辑
     *
     * @param sp
     * @param ticketValidator
     * @param userDetailsService
     * @return
     */
    @Bean
    public CasAuthenticationProvider
casAuthenticationProvider(ServiceProperties sp, TicketValidator
ticketValidator, UserDetailsService userDetailsService) {
        CasAuthenticationProvider provider = new CasAuthenticationProvider();
        provider.setServiceProperties(sp);
        provider.setTicketValidator(ticketValidator);
        provider.setUserDetailsService(userDetailsService);
        provider.setKey("blurooo");
        return provider;
    }


    /**
     *
     * 提供 CAS 验证专用过滤器，过滤器的验证逻辑由 CasAuthenticationProvider 提供
     *
     * @param sp
     * @param ap
     * @return
     */
    @Bean
    public CasAuthenticationFilter casAuthenticationFilter(ServiceProperties
```

```java
sp, AuthenticationProvider ap) {
    CasAuthenticationFilter filter = new CasAuthenticationFilter();
    filter.setServiceProperties(sp);
    filter.setAuthenticationManager(new
ProviderManager(Arrays.asList(ap)));
    return filter;
}

/**
 *
 * 接受CAS服务器发出的注销请求
 *
 * @return
 */
@Bean
public SingleSignOutFilter singleSignOutFilter() {
    SingleSignOutFilter singleSignOutFilter = new SingleSignOutFilter();
    singleSignOutFilter.setCasServerUrlPrefix(casServerPrefix);
    singleSignOutFilter.setIgnoreInitConfiguration(true);
    return singleSignOutFilter;
}

/**
 *
 * 将注销请求转发到CAS Server
 *
 * @return
 */
@Bean
public LogoutFilter logoutFilter() {
    LogoutFilter logoutFilter = new LogoutFilter(casServerLogout, new
SecurityContextLogoutHandler());
    // 设置客户端注销请求的路径
    logoutFilter.setFilterProcessesUrl(casClientLogoutRelative);
    return logoutFilter;
}
}
```

3. 配置 Spring Security

在 Spring Security 中，我们要求/user 下的所有请求都必须拥有 USER 这个角色，并且在内存中初始化了一个用户名 casuser，同样赋予了 USER 角色。也就是说，我们只要使用 casuser

登录 CAS Server，就能得到访问/user 的权限。

```java
@EnableWebSecurity(debug = true)
public class WebSecurityConfig extends WebSecurityConfigurerAdapter {


    @Autowired
    private AuthenticationProvider authenticationProvider;

    @Autowired
    private AuthenticationEntryPoint entryPoint;

    @Autowired
    private SingleSignOutFilter singleSignOutFilter;

    @Autowired
    private LogoutFilter requestSingleLogoutFilter;

    @Autowired
    private CasAuthenticationFilter casAuthenticationFilter;

    @Override
    protected void configure(AuthenticationManagerBuilder auth) {
        auth.authenticationProvider(authenticationProvider);
    }

    @Override
    protected void configure(HttpSecurity http) throws Exception {
        http
            .authorizeRequests()
                .antMatchers("/user/**").hasRole("USER")
                .antMatchers("/login/cas", "/favicon.ico", "/error").permitAll()
                .anyRequest().authenticated()
                .and()
            .exceptionHandling()
                .authenticationEntryPoint(entryPoint)
                .and()
            .addFilter(casAuthenticationFilter)
                .addFilterBefore(singleSignOutFilter, CasAuthenticationFilter.class)
                .addFilterBefore(requestSingleLogoutFilter, LogoutFilter.class);
    }

}
```

第 11 章 HTTP 认证

除系统内维护的用户名和密码认证技术外，Spring Security 还支持 HTTP 层面的认证技术，包括 HTTP 基本认证和 HTTP 摘要认证两种。

11.1 HTTP 基本认证

HTTP 基本认证是在 RFC2616 中定义的一种认证模式，优点是使用简单、没有复杂页面交互。

HTTP 基本认证有 4 个步骤：

（1）客户端发起一条没有携带认证信息的请求。

（2）服务器返回一条 401 Unauthorized 响应，并在 WWW-Authentication 首部说明认证形式，当进行 HTTP 基本认证时，WWW-Authentication 会被设置为 Basic realm="被保护页面"。

（3）客户端收到 401 Unauthorized 响应后，弹出对话框，询问用户名和密码。当用户完成后，客户端将用户名和密码使用冒号拼接并编码为 Base64 形式，然后放入请求的 Authorization 首部发送给服务器。

（4）服务器解码得到客户端发来的用户名和密码，并在验证它们是正确的之后，返回客户端请求的报文。

过程如图 11-1 所示。

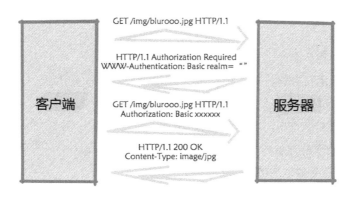

图 11-1

如果不使用浏览器访问 HTTP 基本认证保护的页面，则自行在请求头中设置 Authorization 也是可以的。

总体而言，HTTP 基本认证是一种无状态的认证方式，与表单认证相比，HTTP 基本认证是一种基于 HTTP 层面的认证方式，无法携带 session，即无法实现 Remember-me 功能。另外，用户名和密码在传递时仅做一次简单的 Base64 编码，几乎等同于明文传输，极易出现密码被窃听和重放攻击等安全性问题，在实际系统开发中很少使用这种方式来进行安全验证。如果有必要，也应使用加密的传输层（例如 HTTPS）来保障安全。

11.2 HTTP 摘要认证

HTTP 摘要认证和 HTTP 基本认证一样，也是在 RFC2616 中定义的认证模式，RFC2617 专门对这两种认证模式做了规定。与 HTTP 基本认证相比，HTTP 摘要认证使用对通信双方都可知的口令进行校验，且最终的传输数据并非明文形式。HTTP 摘要基本认证意在解决 HTTP 基本认证存在的大部分严重漏洞，但不应将其认为是 Web 安全的最终解决方案。

11.2.1 认识 HTTP 摘要认证

与 HTTP 基本认证类似，HTTP 摘要认证也是基于简单的"挑战—回应"范例，即在未经验证的请求发起时，服务器会首先返回一个 401 应答（挑战），并携带验证相关的参数，期待客户端依据这些参数继续做出回应，以完成整个验证过程。

HTTP 摘要认证的回应与 HTTP 基本认证相比要复杂得多，下面看看 HTTP 摘要认证中涉及的一些参数：

- username：用户名。
- password：用户密码。
- realm：认证域，由服务器返回。
- opaque：透传字符串，客户端应原样返回。
- method：请求的方法。
- nonce：由服务器生成的随机字符串。
- nc：即 nonce-count，指请求的次数，用于计数，防止重放攻击。qop 被指定时，nc 也必须被指定。
- cnonce：客户端发给服务器的随机字符串，qop 被指定时，cnonce 也必须被指定。
- qop：保护级别，客户端根据此参数指定摘要算法。若取值为 auth，则只进行身份验证；若取值为 auth-int，则还需要校验内容完整性。
- uri：请求的 uri。
- response：客户端根据算法算出的摘要值。
- algorithm：摘要算法，目前仅支持 MD5。
- entity-body：页面实体，非消息实体，仅在 auth-int 中支持。

通常服务器携带的数据包括 realm、opaque、nonce、qop 等字段，如果客户端需要做出验证回应，就必须按照一定的算法计算得到一些新的数据并一起返回。

11.2.2　Spring Security 对 HTTP 摘要认证的集成支持

对于服务器而言，最重要的字段是 nonce；对于客户端而言，最重要的字段是 response。

nonce 是由服务器生成的随机字符串，包含过期时间和密钥。在 Spring Security 中，其生成算法如下。

```
base64(expirationTime + ":" + md5(expirationTime + ":" + key))
```

其中，expirationTime 默认为 300s，在 DigestAuthenticatonEntryPoint 中可以找到 Spring Security 发送"挑战"数据的过程。

```
private int nonceValiditySeconds = 300;

public void commence(HttpServletRequest request, HttpServletResponse response,
AuthenticationException authException) throws IOException, ServletException {
```

```
    long expiryTime = System.currentTimeMillis() +
(long)(this.nonceValiditySeconds * 1000);
    String signatureValue = DigestAuthUtils.md5Hex(expiryTime + ":" + this.key);
    String nonceValue = expiryTime + ":" + signatureValue;
    String nonceValueBase64 = new String(Base64.encode(nonceValue.getBytes()));
    String authenticateHeader = "Digest realm=\"" + this.realmName + "\", 
qop=\"auth\", nonce=\"" + nonceValueBase64 + "\"";
    if(authException instanceof NonceExpiredException) {
        authenticateHeader = authenticateHeader + ", stale=\"true\"";
    }

    if(logger.isDebugEnabled()) {
        logger.debug("WWW-Authenticate header sent to user agent: " + 
authenticateHeader);
    }

    response.addHeader("WWW-Authenticate", authenticateHeader);
    response.sendError(401, authException.getMessage());
}
```

Spring Security 默认实现了 qop 为 auth 的摘要认证模式。如果在客户端最后发起的"回应"中，摘要有效但已过期，那么 Spring Security 会重新发回一个"挑战"，并增加 stale=true 字段告诉客户端不需要重新弹出验证框，用户名和密码是正确的，只需使用新的 nonce 尝试即可。

response 是客户端最关注的字段，它是整个验证能否通过的关键，它的算法取决于 qop，如果 qop 未指定，那么它的算法如下。

```
A1 = md5(username:realm:password)
A2 = md5(method:uri)
response = md5(A1:nonce:A2)
```

如果 qop 指定为 auth，则算法如下。

```
A1 = md5(username:realm:password)
A2 = md5(method:uri)
response = md5(A1:nonce:nc:cnonce:qop:A2)
```

这在 Spring Security 的实现代码中有体现（与 RFC2617 标准相比略有变动）。

```
static String generateDigest(boolean passwordAlreadyEncoded, String username,
String realm, String password, String httpMethod, String uri, String qop, String
nonce, String nc, String cnonce) throws IllegalArgumentException {
    String a2 = httpMethod + ":" + uri;
    String a2Md5 = md5Hex(a2);
```

```
    String a1Md5;
    if(passwordAlreadyEncoded) {
        a1Md5 = password;
    } else {
        a1Md5 = encodePasswordInA1Format(username, realm, password);
    }

    String digest;
    if(qop == null) {
        digest = a1Md5 + ":" + nonce + ":" + a2Md5;
    } else {
        if(!"auth".equals(qop)) {
            throw new IllegalArgumentException("This method does not support a qop: '" + qop + "'");
        }

        digest = a1Md5 + ":" + nonce + ":" + nc + ":" + cnonce + ":" + qop + ":" + a2Md5;
    }

    return md5Hex(digest);
}
```

验证的大体流程是：客户端首先按照约定的算法计算并发送 response，服务器接收之后，以同样的方式计算得到一个 response。如果两个 response 相同，则证明该摘要正确。接着用 base64 解码原 nonce 得到过期时间，以验证该摘要是否还有效。

需要注意的是，由于 HTTP 摘要认证必须读取用户的明文密码，所以不应该在 Spring Security 中使用任何密码加密方式。

11.2.3　编码实现

把 Spring Security 接入 HTTP 摘要认证的过程相对比较麻烦，因为它没有提供配置器，所以我们需要自定义一个 DigestAuthenticationFilter，用于处理该认证。

```
@EnableWebSecurity
public class WebSecurityConfig extends WebSecurityConfigurerAdapter {

    @Autowired
    private DigestAuthenticationEntryPoint myDigestEntryPoint;

    @Autowired
```

```java
    private UserDetailsService userDetailsService;

    // DigestAuthenticationEntryPoint 用于配置HTTP摘要认证部分允许自定义的数据
    @Bean
    public DigestAuthenticationEntryPoint digestEntryPoint() {
        DigestAuthenticationEntryPoint digestAuthenticationEntryPoint = new
DigestAuthenticationEntryPoint();
        digestAuthenticationEntryPoint.setKey("https://github.com/blurooo");
        digestAuthenticationEntryPoint.setRealmName("spring security");
        digestAuthenticationEntryPoint.setNonceValiditySeconds(500);
        return digestAuthenticationEntryPoint;
    }

    // 过滤器指定了 DigestAuthenticationEntryPoint 和 UserDetailsService，这里的
    // UserDetailsService 是必须要指定的，Spring Security 不会主动注入
    public DigestAuthenticationFilter digestAuthenticationFilter() {
        DigestAuthenticationFilter digestFilter = new
DigestAuthenticationFilter();
        digestFilter.setAuthenticationEntryPoint(myDigestEntryPoint);
        digestFilter.setUserDetailsService(userDetailsService);
        return digestFilter;
    }

    @Override
    protected void configure(HttpSecurity http) throws Exception {
        http
        .authorizeRequests()
            .antMatchers("/admin/api/**").hasAuthority("ROLE_ADMIN")
            .antMatchers("/user/api/**").hasRole("USER")
            .antMatchers("/app/api/**").permitAll()
            .anyRequest().authenticated()
            .and()
        .csrf().disable()
        // 当未经认证就访问受保护资源时，会被该认证入口点处理
        .exceptionHandling().authenticationEntryPoint(myDigestEntryPoint)
            .and()
        // 把自定义过滤器加到过滤器链中
        .addFilter(digestAuthenticationFilter());
    }
}
```

通过查看实际的请求交互，可以加深对整个流程的理解。如图 11-2 所示，在浏览器弹出的认证框中，单击"取消"按钮，可以看到该请求的"挑战"内容。

在认证通过之后，刷新页面可以看到该请求的"回应"内容，如图 11-3 所示。

图 11-2

图 11-3

除浏览器自动实现该算法外，XMLHttpRequest 也有支持。

```
XMLHttpRequest.open(method, url, async, username, password)
```

可以在浏览器控制台实验，如图 11-4 所示。

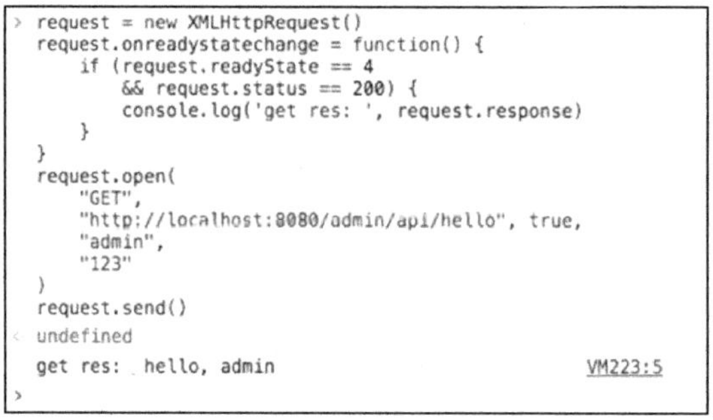

图 11-4

HTTP 摘要认证与 HTTP 基本认证一样，都是基于 HTTP 层面的认证方式，不使用 session，因而不支持 Remember-me。虽然解决了 HTTP 基本认证密码明文传输的问题，但并未解决密码明文存储的问题，依然存在安全隐患。HTTP 摘要认证与 HTTP 基本认证相比，仅仅在非加密的传输层中有安全优势，但是其相对复杂的实现流程，使得它并不能成为一种被广泛使用的认证方式。

第 12 章
@EnableWebSecurity与过滤器链机制

为什么加上@EnableWebSecurity 注解就可以让 Spring Security 起作用？Spring Security 又是通过什么方式来拦截请求并执行认证的？下面就带着这两个问题，深入源码一探究竟。

12.1 @EnableWebSecurity

@EnableWebSecurity 是开启 Spring Security 的默认行为，它通过@Import 注解导入了 WebSecurityConfiguration 类。即当我们使用 @EnableWebSecurity 注解时，等同于将 WebSecurityConfiguration 类放入 Spring 的 IoC 容器里。也就是说，WebSecurityConfiguration 借由@EnableWebSecurity 注解得到初始化的机会。

```
@Retention(RetentionPolicy.RUNTIME)
@Target({ElementType.TYPE})
@Documented
@Import({WebSecurityConfiguration.class, SpringWebMvcImportSelector.class})
@EnableGlobalAuthentication
@Configuration
public @interface EnableWebSecurity {
    boolean debug() default false;
}
```

@EnableWebSecurity 还有一个 debug 参数用于指定是否采用调试模式，默认为 false。在调试模式下，每个请求的详细信息和所经过的过滤器，甚至其调用栈都会被打印到控制台。

```
***********************************************************
Request received for GET '/user/api/hello':

org.springframework.session.web.http.SessionRepositoryFilter$SessionRepositoryRequestWrapper@69439a45

servletPath:/user/api/hello
pathInfo:null
headers:
host: localhost:8081
connection: keep-alive
upgrade-insecure-requests: 1
user-agent: Mozilla/5.0 (Macintosh; Intel Mac OS X 10_12_4) AppleWebKit/537.36 (KHTML, like Gecko) Chrome/64.0.3282.186 Safari/537.36
accept: text/html,application/xhtml+xml,application/xml;q=0.9,image/webp,image/apng,*/*;q=0.8
accept-encoding: gzip, deflate, br
accept-language: zh-CN,zh;q=0.9,en;q=0.8
cookie: io=d2LRCbg-jYfwZWXXAAAA; SESSION=396509a0-f1cf-49f4-9225-e18fc5307d68
key: kjJSK123hSskAsSSsdjhA12UI1IUDH12
```

12.2　WebSecurityConfiguration

WebSecurityConfiguration 用于初始化 WebSecurity 配置。首先，在 setFilterChainProxySecurityConfigurer 方法中，它以配置 Spring Security 时继承自 WebSecurityConfigurerAdapter 的配置类来初始化一个 SecurityConfigurer 列表。Spring Security 以 SecurityConfigurer 列表为依据，启用所需的安全策略。

```java
// webSecurityConfigurers 继承了 WebSecurityConfigurerAdapter 的配置器（允许配置多个）
@Autowired(required = false)
public void setFilterChainProxySecurityConfigurer(ObjectPostProcessor<Object> objectPostProcessor,
        @Value("#{@autowiredWebSecurityConfigurersIgnoreParents.getWebSecurityConfigurers()}") List<SecurityConfigurer<Filter, WebSecurity>> webSecurityConfigurers) throws Exception {
    this.webSecurity = (WebSecurity)objectPostProcessor.postProcess(new WebSecurity(objectPostProcessor));
    if(this.debugEnabled != null) {
        this.webSecurity.debug(this.debugEnabled.booleanValue());
    }

    Collections.sort(webSecurityConfigurers, WebSecurityConfiguration.AnnotationAwareOrderComparator.INSTANCE);
    Integer previousOrder = null;
    Object previousConfig = null;

    Iterator var5;
    SecurityConfigurer config;
    for(var5 = webSecurityConfigurers.iterator(); var5.hasNext(); previousConfig = config) {
        config = (SecurityConfigurer)var5.next();
        Integer order =
```

```
Integer.valueOf(WebSecurityConfiguration.AnnotationAwareOrderComparator.look
upOrder(config));
        if(previousOrder != null && previousOrder.equals(order)) {
            throw new IllegalStateException("@Order on WebSecurityConfigurers
must be unique. Order of " + order + " was already used on " + previousConfig
+ ", so it cannot be used on " + config + " too.");
        }

        previousOrder = order;
    }

    var5 = webSecurityConfigurers.iterator();

    while(var5.hasNext()) {
        config = (SecurityConfigurer)var5.next();
        // 将配置的每一个SecurityConfigurer列表传递给WebSecurity
        this.webSecurity.apply(config);
    }

    this.webSecurityConfigurers = webSecurityConfigurers;
}
```

接着构建过滤器链。

```
// 提供一个名为springSecurityFilterChain的bean，返回一个Filter对象
@Bean(name = {"springSecurityFilterChain"})
public Filter springSecurityFilterChain() throws Exception {
    boolean hasConfigurers = this.webSecurityConfigurers != null
&& !this.webSecurityConfigurers.isEmpty();
    if(!hasConfigurers) {
        // 如果没有配置过Spring Security，则会以WebSecurityConfigurerAdapter中的
        //配置作为默认行为
        WebSecurityConfigurerAdapter adapter =
(WebSecurityConfigurerAdapter)this.objectObjectPostProcessor.postProcess(new
WebSecurityConfigurerAdapter() {
        });
        this.webSecurity.apply(adapter);
    }

    return (Filter)this.webSecurity.build();
}
```

Spring Security 的核心实现是通过一条过滤器链来确定用户的每一个请求应该得到什么样的反馈，如图 12-1 所示。

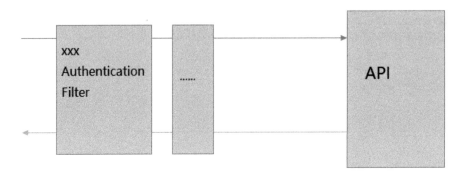

图 12-1

为了让 Spring Security 生效，其中一个重要的环节是在 web.xml 中配置 Spring Security 提供的过滤器。

```
<filter>
   <filter-name>springSecurityFilterChain</filter-name>
<filter-class>org.springframework.web.filter.DelegatingFilterProxy</filter-class>
</filter>

<filter-mapping>
   <filter-name>springSecurityFilterChain</filter-name>
   <url-pattern>/*</url-pattern>
</filter-mapping>
```

DelegatingFilterProxy 是 Spring 提供的一个标准的 Servlet Filter 代理，简单来说，上面的配置等同于注入一个名为 springSecurityFilterChain 的 bean，并代理该 bean 提供的过滤器。也就是说，在这个配置中，最终起作用的过滤器是什么完全取决于 springSecurityFilterChain。我们在之前的实践中没有配置 web.xml 也能正常使用 Spring Security，是因为 Spring Boot 已经自动配置了 web.xml。

```
@Configuration
@ConditionalOnWebApplication
@EnableConfigurationProperties
@ConditionalOnClass({AbstractSecurityWebApplicationInitializer.class,
SessionCreationPolicy.class})
@AutoConfigureAfter({SecurityAutoConfiguration.class})
public class SecurityFilterAutoConfiguration {
    private static final String DEFAULT_FILTER_NAME =
```

```
"springSecurityFilterChain";

    public SecurityFilterAutoConfiguration() {
    }

    @Bean
    @ConditionalOnBean(
        name = {"springSecurityFilterChain"}
    )
    public DelegatingFilterProxyRegistrationBean
securityFilterChainRegistration(SecurityProperties securityProperties) {
        DelegatingFilterProxyRegistrationBean registration = new
DelegatingFilterProxyRegistrationBean("springSecurityFilterChain", new
ServletRegistrationBean[0]);
        registration.setOrder(securityProperties.getFilterOrder());

registration.setDispatcherTypes(this.getDispatcherTypes(securityProperties));
        return registration;
    }

    ...
}
```

前面提过，Spring Security 是由过滤器链实现的，为什么在配置时却只有 springSecurityFilterChain 一个过滤器呢？下面来揭开 springSecurityFilterChain 这个过滤器的真面目。

```
@Bean(name = {"springSecurityFilterChain"})
public Filter springSecurityFilterChain() throws Exception {

    ...

    // 通过调用 WebSecurity 的 build 方法生成过滤器
    return (Filter)this.webSecurity.build();
}
```

WebSecurity 的 build 方法最终调用的是 doBuild（WebSecurity 继承了 AbstractConfiguredSecurityBuilder，doBuild 正是由它提供的）。

```
public final O build() throws Exception {
    if(this.building.compareAndSet(false, true)) {
        this.object = this.doBuild();
        return this.object;
    } else {
        throw new AlreadyBuiltException("This object has already been built");
```

doBuild 调用的是 WebSecurity 的 performBuild 方法。

```
protected final O doBuild() throws Exception {
    LinkedHashMap var1 = this.configurers;
    synchronized(this.configurers) {
        // 按照状态依次执行相应方法
        this.buildState =
AbstractConfiguredSecurityBuilder.BuildState.INITIALIZING;
        this.beforeInit();
        // 初始化状态,通过调用 WebSecurityConfigurerAdapter 的 init,将所有
        // HttpSecurity 添加到 WebSecurity 里
        this.init();
        this.buildState =
AbstractConfiguredSecurityBuilder.BuildState.CONFIGURING;
        this.beforeConfigure();
        this.configure();
        this.buildState =
AbstractConfiguredSecurityBuilder.BuildState.BUILDING;
        // 在 BUILDING 阶段调用 WebSecurity 的 performBuild 方法
        O result = this.performBuild();
        this.buildState = AbstractConfiguredSecurityBuilder.BuildState.BUILT;
        return result;
    }
}
```

在 performBuild 方法里,Spring Security 完成了所有过滤器的构建,最终返回一个过滤器链代理类 filterChainProxy。

```
protected Filter performBuild() throws Exception {
    Assert.state(!this.securityFilterChainBuilders.isEmpty(), "At least one SecurityBuilder<? extends SecurityFilterChain> needs to be specified. Typically this done by adding a @Configuration that extends WebSecurityConfigurerAdapter. More advanced users can invoke " + WebSecurity.class.getSimpleName() + ".addSecurityFilterChainBuilder directly");
    int chainSize = this.ignoredRequests.size() + this.securityFilterChainBuilders.size();
    List<SecurityFilterChain> securityFilterChains = new ArrayList(chainSize);
    Iterator var3 = this.ignoredRequests.iterator();

    while(var3.hasNext()) {
        RequestMatcher ignoredRequest = (RequestMatcher)var3.next();
        securityFilterChains.add(new DefaultSecurityFilterChain(ignoredRequest,
```

```
            new Filter[0]));
        }

        var3 = this.securityFilterChainBuilders.iterator();

        while(var3.hasNext()) {
            SecurityBuilder<? extends SecurityFilterChain>
                securityFilterChainBuilder = (SecurityBuilder)var3.next();
            // 简单来说，就是每一个HttpSecurity生成一条过滤器链，HttpSecurity则来自我们配置
            // 的WebSecurityConfigure（继承WebSecurityConfigureAdapter）
            securityFilterChains.add(securityFilterChainBuilder.build());
        }

        // 过滤器链实际是被filterChainProxy代理的
        FilterChainProxy filterChainProxy = new
FilterChainProxy(securityFilterChains);
        if(this.httpFirewall != null) {
            filterChainProxy.setFirewall(this.httpFirewall);
        }

        filterChainProxy.afterPropertiesSet();
        Filter result = filterChainProxy;
        if(this.debugEnabled) {
this.logger.warn("\n\n********************************************************
**************\n**********          Security debugging is enabled.
**************\n**********    This may include sensitive information.
**************\n**********         Do not use in a production system!
**************\n********************************************************
*******\n\n");
            result = new DebugFilter(filterChainProxy);
        }

        this.postBuildAction.run();
        return (Filter)result;
    }
```

filterChainProxy 间接继承了 Filter，可以作为真正的过滤器使用。它会携带若干条过滤器链，并在承担过滤器职责时，将其派发到所有过滤器链的每一个过滤器上。

```
public void doFilter(ServletRequest request, ServletResponse response,
FilterChain chain) throws IOException, ServletException {
    boolean clearContext = request.getAttribute(FILTER_APPLIED) == null;
    if(clearContext) {
        try {
```

```
            request.setAttribute(FILTER_APPLIED, Boolean.TRUE);
            // 派发到过滤器链上
            this.doFilterInternal(request, response, chain);
        } finally {
            SecurityContextHolder.clearContext();
            request.removeAttribute(FILTER_APPLIED);
        }
    } else {
        this.doFilterInternal(request, response, chain);
    }
}
```

doFilterInternal 是真正执行虚拟过滤器链逻辑的方法。

```
private void doFilterInternal(ServletRequest request, ServletResponse response,
FilterChain chain) throws IOException, ServletException {
    // 附上 Spring Security 提供的 HTTP 防火墙
    FirewalledRequest fwRequest =
this.firewall.getFirewalledRequest((HttpServletRequest)request);
    HttpServletResponse fwResponse =
this.firewall.getFirewalledResponse((HttpServletResponse)response);
    // 按照配置的 RequestMatcher, 决定每一个请求会经过哪些过滤器
    List<Filter> filters = this.getFilters((HttpServletRequest)fwRequest);
    if(filters != null && filters.size() != 0) {
        // 所有过滤器合并成一条虚拟过滤器链
        FilterChainProxy.VirtualFilterChain vfc = new
FilterChainProxy.VirtualFilterChain(fwRequest, chain, filters);
        // 模拟过滤器的执行流程, 执行整条过滤器链
        vfc.doFilter(fwRequest, fwResponse);
    } else {
        if(logger.isDebugEnabled()) {
            logger.debug(UrlUtils.buildRequestUrl(fwRequest) + (filters ==
null?" has no matching filters":" has an empty filter list"));
        }

        fwRequest.reset();
        chain.doFilter(fwRequest, fwResponse);
    }
}

private static class VirtualFilterChain implements FilterChain {
    private final FilterChain originalChain;
    private final List<Filter> additionalFilters;
```

```java
    private final FirewalledRequest firewalledRequest;
    private final int size;
    private int currentPosition;

    private VirtualFilterChain(FirewalledRequest firewalledRequest,
        FilterChain chain, List<Filter> additionalFilters) {
        this.currentPosition = 0;
        this.originalChain = chain;
        this.additionalFilters = additionalFilters;
        this.size = additionalFilters.size();
        this.firewalledRequest = firewalledRequest;
    }

    public void doFilter(ServletRequest request, ServletResponse response)
throws IOException, ServletException {
        if(this.currentPosition == this.size) {
            if(FilterChainProxy.logger.isDebugEnabled()) {

FilterChainProxy.logger.debug(UrlUtils.buildRequestUrl(this.firewalledReques
t) + " reached end of additional filter chain; proceeding with original chain");
            }

            this.firewalledRequest.reset();

            // 执行过滤器链后，调用真实的 FilterChain，完成原生过滤器的剩余逻辑
            this.originalChain.doFilter(request, response);
        } else {
            ++this.currentPosition;
            Filter nextFilter =
(Filter)this.additionalFilters.get(this.currentPosition - 1);
            if(FilterChainProxy.logger.isDebugEnabled()) {

FilterChainProxy.logger.debug(UrlUtils.buildRequestUrl(this.firewalledReques
t) + " at position " + this.currentPosition + " of " + this.size + " in additional
filter chain; firing Filter: '" + nextFilter.getClass().getSimpleName() + "'");
            }
            // 通过改变下标回调的方式按照顺序执行每一个过滤器
            nextFilter.doFilter(request, response, this);
        }

    }
}
```

第 3 部分

第 13 章

用Spring Social实现OAuth对接

OAuth 解决了在用户不提供密码给第三方应用的情况下，让第三方应用有权获取用户数据以及基本信息的难题。

13.1 OAuth 简介

13.1.1 什么是 OAuth

开放授权（Open Authorization，OAuth）是一种资源提供商用于授权第三方应用代表资源所有者获取有限访问权限的授权机制。由于在整个授权过程中，第三方应用都无须触及用户的密码就可以取得部分资源的使用权限，所以 OAuth 是安全开放的。

OAuth 第一个版本诞生于 2007 年 12 月，并于 2010 年 4 月正式被 IETF 作为标准发布（编号 RFC 5849）。由于 OAuth1.0 复杂的签名逻辑以及单一的授权流程存在较大缺陷，随后标准工作组又推出了 OAuth2.0 草案，并在 2012 年 10 月正式发布其标准（编号 RFC 6749）。OAuth2.0 放弃了 OAuth1.0 中让开发者感到痛苦的数字签名和加密方案，使用已经得到验证并广泛使用的 HTTPS 技术作为安全保障手段。OAuth2.0 与 OAuth1.0 互不兼容，由于 OAuth1.0 已经基本退出历史舞台，所以下面提到的 OAuth 都是指 OAuth2.0。

说到 OAuth，相信很多人都曾不知不觉使用过。例如，在登录 CSDN 网站时，有 4 种第三方登录渠道可供选择，如图 13-1 所示。

从左到右分别是 QQ、新浪微博、百度、开源中国和 GitHub。当尝试使用 QQ 登录时，就会跳转到 QQ 授权登录页面（该页面位于 QQ 站点，在登录成功之后再跳转回 CSDN 网站，从而避免在第三方站点直接提交 QQ 密码），如图 13-2 所示。

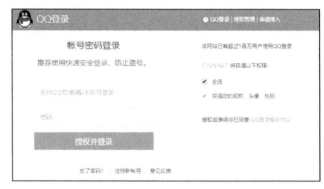

图 13-1　　　　　　　　　　　　　　图 13-2

右边勾选的权限仅有用户的基本信息，包括昵称、头像和性别。也就是说，通过 QQ 账号登录 CSDN 网站之后，CSDN 网站只能获取这一部分权限，其他隐私数据则是完全不可见的。除此之外，我们还可以单击右上角的"授权管理"对已经授权过的站点进行权限管理，包括"获取移动支付相关信息"和"取消全部授权"，如图 13-3 所示。

图 13-3

从 QQ 授权登录这一系列机制中，我们基本可以窥探到 OAuth 认证的流程和形式。

13.1.2 OAuth 的运行流程

想要理解 OAuth 的运行流程，则必须要认识 4 个重要的角色。

（1）Resource Owner：资源所有者，通常指用户，例如每一个 QQ 用户。

（2）Resource Server：资源服务器，指存放用户受保护资源的服务器，通常需要通过 Access Token（访问令牌）才能进行访问。例如，存储 QQ 用户基本信息的服务器，充当的便是资源服务器的角色。

（3）Client：客户端，指需要获取用户资源的第三方应用，如 CSDN 网站。

（4）Authorization Server：授权服务器，用于验证资源所有者，并在验证成功之后向客户端发放相关访问令牌。例如，QQ 授权登录页面。

图 13-4 很清晰地描述了这 4 种角色是如何进行交互的。

图 13-4

（A）客户端要求用户提供授权许可。

（B）用户同意向客户端提供授权许可。

（C）客户端携带用户提供的授权许可向授权服务器申请资源服务器的访问令牌。

（D）授权服务器验证客户端及其携带的授权许可，确认有效后发放访问令牌。

（E）客户端使用访问令牌向资源服务器申请资源。

（F）资源服务器验证访问令牌，确认无误后向客户端提供资源。

在这个流程中，B 步骤是最为关键的一步，OAuth 定义了 4 种授权模式，用于将用户的授权许可提供给客户端。

1. 授权码模式（Authorization Code）

授权码模式是功能最完整、流程最严密的授权模式，它将用户引导到授权服务器进行身份验证，授权服务器将发放的访问令牌传递给客户端。例如，前面介绍的 QQ 登录方式，就是由 CSDN 网站引导到 QQ 授权服务器进行身份验证的。一个典型的 QQ 登录页面 URL 如下。

```
https://graph.qq.com/oauth2.0/show?which=Login&display=pc&response_type=code
&client_id=100270989&redirect_uri=https://passport.csdn.net/account/login?oa
uth_provider=QQProvider&state=test
```

其中，response_type 指授权类型，为必要项，固定为 code。client_id 指客户端 id，为必要项。

state 指客户端的状态，通常在授权服务器重定向时原样返回。

scope 为申请的权限范围，如获取用户信息、获取用户相册等，由授权服务器抽象为具体的条目。

redirect_uri 为授权通过后的重定向 URL，授权服务器将在用户登录完成之后重定向到类似下面的地址。

```
https://passport.csdn.net/account/login?oauth_provider=QQProvider&code=xxx&s
tate=test
```

code 为申请访问令牌必备的授权码（有效期较短，注意与访问令牌的区别）。客户端拿到 code 之后需要向授权服务器申请访问令牌（仅可使用一次，用完作废）。

申请令牌时也有一些关键参数，其中，grant_type 指授权类型，在授权码模式中，该值需要设置为 authorization_code；client_id 指客户端 id；code 指前面获取的授权码；redirect_uri 指重定向 URL。通过构建一个 HTTP 请求发起访问令牌的申请，如果成功，则会得到访问令牌，以及一些当令牌刷新时需要的参数。

```
https://graph.qq.com/oauth2.0/xxx?grant_type=authorization_code&code=xxx&...
```

授权码模式的客户端指接入 OAuth 的第三方应用，例如，CSDN 网站就属于一个客户端。

授权码模式的完整运行流程如图 13-5 所示。

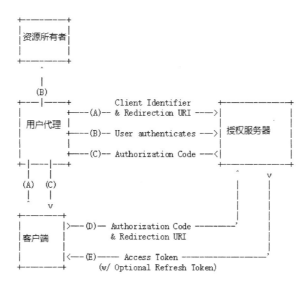

图 13-5

2．隐式授权模式（Implicit）

隐式授权模式的客户端一般是指用户浏览器。访问令牌通过重定向的方式传递到用户浏览器中，再通过浏览器的 JavaScript 代码来获取访问令牌。由于访问令牌直接暴露在浏览器端，所以隐式授权模式可能会导致访问令牌被黑客获取，仅适用于需要临时访问的场景。当然，每一种授权模式都有其应用场景。例如，QQ 的 OAuth 针对移动端用户采用的便是隐式授权模式。

隐式授权模式的完整运行流程如图 13-6 所示。

与授权码模式相比，用户的登录环节是一样的，只是在授权成功之后的重定向上，授权码模式是携带一个认证码，由客户端（第三方应用后端程序）通过认证码申请访问令牌的；而隐式授权模式则直接将访问令牌作为 URL 的散列部分传递给浏览器。

```
http://graph.qq.com/demo/index.jsp?#access_token=FE04***********************
*CCE2&expires_in=7776000&state=test
```

URL 中#后面的部分被称为 URL HASH，散列部分是专门用于指导浏览器行为的。例如，浏览器页面定位的锚点是通过散列属性判定的，在实际的 HTTP 请求中不会携带该散列部分，所以散列属性并不会传递到第三方应用的后端。

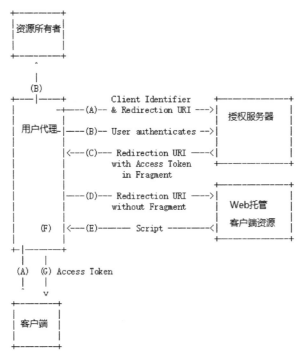

图 13-6

在 OAuth 标准实现中，隐式授权模式在重定向时也会携带若干个参数，包括 access_token，即最关键的访问令牌；expire_in，指该访问令牌在多少秒后过期；state，指客户端的状态参数。

客户端页面可以使用 JavaScript 获取该散列值。

```
var hash = window.location.hash
```

最好利用浏览器的 cookie 储存访问令牌。

3. 密码授权模式（Password Credentials）

顾名思义，就是客户端直接携带用户的密码向授权服务器申请令牌。这种登录操作不再像前两种授权模式一样跳转到授权服务器进行，而是由客户端提供专用页面。如果用户信任该客户端（通常为信誉度高的著名公司），用户便可以直接提供密码，客户端在不储存用户密码的前提下完成令牌的申请。

密码授权模式的完整运行流程如图 13-7 所示。

当然，除非前两种授权模式无法顺利进行，否则不考虑使用密码授权模式。

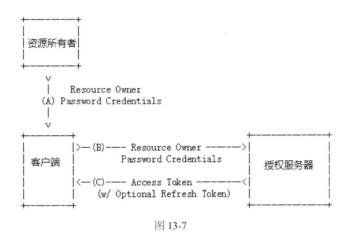

图 13-7

4．客户端授权模式（Client Credentials）

客户端授权模式实际上并不属于 OAuth 的范畴，因为它的关注点不再是用户的私有信息或数据，而是一些由资源服务器持有但并非完全公开的数据，如微信的公众平台授权等。

客户端授权模式通常由客户端提前向授权服务器申请应用公钥、密钥，并通过这些关键信息向授权服务器申请访问令牌，从而得到资源服务器提供的资源。客户端授权模式的完整运行流程如图 13-8 所示。

图 13-8

OAuth 仅仅定义了一个粗略的流程规范，具体的实现细节由实际的授权服务器制定。

总体而言，不管是作为客户端接入知名 OAuth 服务提供商，还是使用 OAuth 保护自身资源，如果关注 Web 服务安全，那么 OAuth 是绕不开的。

13.2　QQ 互联对接准备

在接入一个新的 OAuth 服务之前，首先需要申请对应服务提供商的 OAuth 应用，然后了解其 API 交互。

13.2.1 申请 QQ 互联应用

如果第三方应用需要使用 QQ 登录，那么必须通过 QQ 互联申请接入。QQ 互联的地址为 https://connect.qq.com/。首次接入时需要填写认证信息，既可以选择企业接入，也可以选择个人接入，提交相对应的资料等待审核即可，如图 13-9 所示。

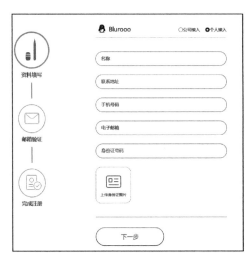

图 13-9

认证完成即可创建应用，如图 13-10 所示。

图 13-10

在此之前需要准备相关已通过备案的域名,具体注意项如下:

◎ 网站域名直接填写域名主体,例如 csdn.net。
◎ 网站回调域填写完整的回调地址,例如,https://blog.csdn.net/oauth/qq(该地址仅为举例说明使用,并非实际存在,此处并非必须使用 HTTPS,OAuth 的安全保障手段虽然是基于 HTTPS 的,但第三方应用不受此限制。另外,该地址不必花费心思准备,因为后面可以随时更改,先填写一个能通过审核的形式即可)。
◎ 提供方填写域名备案时的网站负责人,个人开发者通常为姓名。
◎ 网站备案号必须提供,否则无法通过审核。
◎ 网站应用图标必须提供,否则无法通过审核。

创建并通过审核之后,最终会得到接入的 appId 和 appKey,查看应用接口,可以得到当前拥有的权限,如图 13-11 所示。个人开发者通常只能获取用户基本信息,更多权限需要以企业认证的方式接入。

图 13-11

13.2.2　QQ 互联指南

QQ 互联提供了一份官方指南:http://wiki.connect.qq.com/,基本的接入流程在这份指南上都有提及,但有少部分交互细节没有明确指出,下面将对这部分内容进行总结和分析。

1. 准备工作

◎ 申请 QQ 互联的 appId 和 appKey。
◎ 确保连接畅通,服务器可以正常访问 https://graph.qq.com。

2. 获取 OAuth 相关 API

QQ 提供的 API 在交互上相对混乱，其响应类型为 text/html，响应内容有普通文本、JSONP、JSON 字符串等多种类型。

例如，当成功获取 accessToken 的 API 时会返回类似于下面的结果：

```
access_token=FE04************************CCE2&expires_in=7776000&refresh_tok
en=88E4************************BE14
```

若出现错误，则返回一段 JSONP 形式包装的 JSON 字符串。对响应结果进行解析时，需要着重注意这部分差异。

（1）获取 Authorization Code。

- 请求地址：https://graph.qq.com/oauth2.0/authorize。
- 请求方式：GET。
- 请求参数如图 13-12 所示。

参数	是否必需	含义
response_type	是	授权类型，此值固定为 "code"
client_id	是	申请QQ登录成功后，分配给应用的appid
redirect_uri	是	成功授权后的回调地址，必须是注册appid时填写的主域名下的地址，建议设置为网站首页或网站的用户中心。注意需要将url进行URLEncode
state	是	client端的状态值。用于第三方应用防止CSRF攻击，成功授权后回调时会原样带回。请务必严格按照流程检查用户与state参数状态的绑定
scope	可选	请求用户授权时向用户显示的可进行授权的列表。可填写的值是API文档中列出的接口，以及一些动作型的授权（目前仅有：do_like），如果要填写多个接口名称，请用逗号隔开。例如：scope=get_user_info,list_album,upload_pic,do_like 不传则默认请求对接口get_user_info进行授权。建议控制授权项的数量，只传入必要的接口名称，因为授权项越多，用户越可能拒绝进行任何授权
display	可选	仅PC网站接入时使用。用于展示的样式。不传则默认展示为PC下的样式。如果传入 "mobile"，则展示为mobile端下的样式

图 13-12

- 请求示例：

https://graph.qq.com/oauth2.0/authorize?clientid=101476839&responsetype=code&redirect_uri=http://oauth.chenmuxin.cn:8080/oauth/qq&state=test

◎ 请求效果：跳转至如图 13-13 所示的 QQ 登录页面。

图 13-13

登录成功时将跳转回 redirect_uri 指明的地址，并以参数的形式携带 code。

（2）获取 access_token。

◎ 请求地址：https://graph.qq.com/oauth2.0/token。
◎ 请求方式：GET / POST。
◎ 请求参数如图 13-14 所示。

参数	是否必需	含义
grant_type	是	授权类型，在本步骤中，此值为 "authorization_code"
client_id	是	申请QQ登录成功后，分配给网站的appid
client_secret	是	申请QQ登录成功后，分配给网站的appkey
code	是	上一步返回的authorization code。如果用户成功登录并授权，则会跳转到指定的回调地址，并在URL中带上Authorization Code。例如，回调地址为www.qq.com/my.php，则跳转到：http://www.qq.com/my.php?code=520DD95263C1CFEA087******注意此code会在10分钟内过期
redirect_uri	是	与上面一步中传入的redirect_uri保持一致

图 13-14

◎ 响应结果如下。

```
access_token=FE04************************CCE2&expires_in=7776000&refresh_tok
en=88E4***********************BE14
```

（3）获取 OpenId。

- 请求地址：https://graph.qq.com/oauth2.0/me。
- 请求方式：GET。
- 请求参数如图 13-15 所示。

参数	是否必需	含义
access_token	是	在Step1中获取到的access token

图 13-15

- 响应结果：

```
callback( {"client_id":"YOUR_APPID","openid":"YOUR_OPENID"} );
```

（4）获取用户信息。

- 请求地址：https://graph.qq.com/user/getuserinfo。
- 请求方式：GET。
- 请求参数如图 13-16 所示。

参数	含义
access_token	可通过使用Authorization_Code获取 access_token access_token有3个月有效期
oauth_consumer_key	申请QQ登录成功后，分配给应用的appid
openid	用户的ID，与QQ号码——对应。 可通过调用https://graph.qq.com/oauth2.0/me?access_token=YOUR_ACCESS_TOKEN 来获取

图 13-16

响应结果如下。

```
{
  "ret":0,
  "msg":"",
  "nickname":"Peter",

"figureurl":"http://qzapp.qlogo.cn/qzapp/111111/942FEA70050EEAFBD4DCE2C1FC775E56/30",

"figureurl_1":"http://qzapp.qlogo.cn/qzapp/111111/942FEA70050EEAFBD4DCE2C1FC775E56/50",
```

```
"figureurl_2":"http://qzapp.qlogo.cn/qzapp/111111/942FEA70050EEAFBD4DCE2C1FC
775E56/100",

"figureurl_qq_1":"http://q.qlogo.cn/qqapp/100312990/DE1931D5330620DBD07FB4A5
422917B6/40",

"figureurl_qq_2":"http://q.qlogo.cn/qqapp/100312990/DE1931D5330620DBD07FB4A5
422917B6/100",
    "gender":"男",
    "is_yellow_vip":"1",
    "vip":"1",
    "yellow_vip_level":"7",
    "level":"7",
    "is_yellow_year_vip":"1"
}
```

13.2.3　回调域名准备

创建 QQ 互联应用之后,如果希望在本地进行测试开发,那么可以将该域名映射到本地某个端口上。此处笔者推荐使用 ngrok 服务:ngrok.cc。该服务提供的免费版本支持绑定自定义域名,对于个人开发者而言,基本可以满足需求(ngrok.com 则需要付费才能绑定域名)。

注册完成之后,在用户界面单击"隧道管理→开通隧道→开通免费版本服务器",即可打开如图 13-17 所示页面。

图 13-17

隧道协议选择 http，隧道名称随意填写即可。前置域名为最终映射到公网的访问路径前缀，只要没有被其他人注册都是允许的。由于后面会绑定自定义域名，所以此处的填写内容无关紧要。本地端口可以根据本地运行服务的端口而定。例如，在本地 8080 端口运行测试，就可以设置为 127.0.0.1:8080。

添加完成之后，进入"隧道管理"可以看到刚才开通的隧道，将其修改为自定义域名（自定义域名需要使用 cname 的方式指向 free.ngrok.cc），如图 13-18 所示。

图 13-18

回到隧道管理页面，下载对应客户端，如图 13-19 所示。

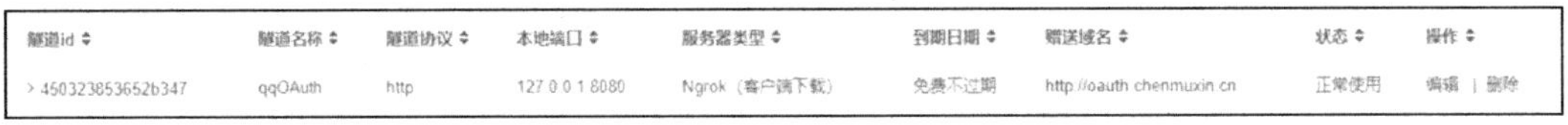

图 13-19

最终，通过客户端运行隧道 id 即可成功完成本地端口映射，如图 13-20 所示。

如果此时本地有服务运行在 8080 端口，那么外网就可以通过自定义的域名进行访问。

虽然这么做是可行的，但实际上 OAuth 服务并不需要这么复杂，因为 OAuth 不像微信公众号一样需要验证回调地址，所以并不需要提供一个能被公网访问的 URL。只要确保地址是备案过的，在开发时，将该地址以 hosts 的形式在本地映射即可。

127.0.0.1 blog.csdn.net

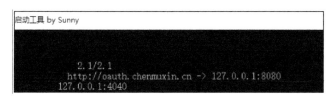

图 13-20

修改 hosts 的方式很简单，Windows 系统的 hosts 文件存放于 C:\windows\system32\drivers\etc\hosts，而 mac 系统的 hosts 文件存放于/etc/hosts，只要用合适的编辑器打开并加入上面的地址即可，之后再访问 blog.csdn.net 时就会被解析到本地。

除可以进行本地映射域名外，甚至不需要在 8080 端口进行开发（在 UNIX 系统中，1024 以下端口需要有管理员权限才能开启），对本地开发测试的阻碍非常小。

13.3　实现 QQ 快捷登录

在 Spring Security 中，如果需要接入 QQ 之类的 OAuth 服务，则可以使用专门用于连接社交平台，实现 OAuth 服务共享的 Spring Social 框架。下面将初步介绍整个接入流程。

13.3.1　引入 Spring Social

Spring Social 是一个专门用于连接社交平台，实现 OAuth 服务共享的框架。我们所熟知的社交平台，如 Facebook、Twitter、微信、新浪微博等，都提供了 OAuth 服务。

Spring Social 有 4 个核心模块：

（1）spring-social-core：提供核心的 OAuth 客户端支持。

（2）spring-social-config：提供 Java 或 XML 配置支持。

（3）spring-social-security：提供 Spring Security 的集成支持。

（4）spring-social-web：提供 Web 环境的连接管理，处理 Web 应用程序和服务提供者之间的来回授权握手。

通常情况下，建议将 4 个模块一并引入，除了 core，其他都是可插拔的，但其他模块同样有使用的必要（如果不是 Spring Security 环境，则不需要引入 Spring Security 模块，但需要引入额外的加密库用于保障安全）。

在笔者成书时，Spring Social 2.0 处于里程碑版本，总体而言，2.0 以后的版本变动不会太大。

引入 Spring Social 与引入其他 Maven 库没有太大的差别，只是里程碑版本需要额外指明其存储库。

```xml
<!-- 由于正式版还没有上架 Maven 官方库，所以里程碑版本应该为其指明存储库，否则无法引入 -->
<repositories>
        <repository>
                <id>org.springframework.maven.milestone</id>
                <name>Spring Maven Milestone Repository</name>
                <url>http://repo.spring.io/milestone</url>
        </repository>
</repositories>
<dependencies>

    <!-- ... -->

        <dependency>
                <groupId>org.springframework.social</groupId>
                <artifactId>spring-social-core</artifactId>
                <version>2.0.0.M4</version>
        </dependency>
        <dependency>
                <groupId>org.springframework.social</groupId>
                <artifactId>spring-social-web</artifactId>
                <version>2.0.0.M4</version>
        </dependency>
        <dependency>
                <groupId>org.springframework.social</groupId>
                <artifactId>spring-social-config</artifactId>
                <version>2.0.0.M4</version>
        </dependency>
        <dependency>
                <groupId>org.springframework.social</groupId>
                <artifactId>spring-social-security</artifactId>
                <version>2.0.0.M4</version>
        </dependency>
</dependencies>
```

Spring Social 为我们封装了 OAuth 的标准认证流程，但接入 OAuth 的目的是为了获取 OAuth 服务提供商提供的数据。不同的 OAuth 服务，资源也不相同，所以我们还需要在 Spring Social

的基础上开发这部分功能。对于大部分全球知名社交平台，Spring 官方都提供了直接支持，如 Facebook、GitHub 等，我们可以在 Spring 官方的 GitHub 账号下搜索 Spring Social，或者直接到 Spring Social 官网查看，即可看到已经支持的各个 Spring Social 子项目。

13.3.2　新增 OAuth 服务支持的流程

Spring Social 可以轻松添加框架尚未支持的 OAuth 服务，如果翻看官方对于一些热门 OAuth 服务的实现库，如 spring-social-twitter、spring-social-facebook 等，官方推荐的代码组织形式如图 13-21 所示。

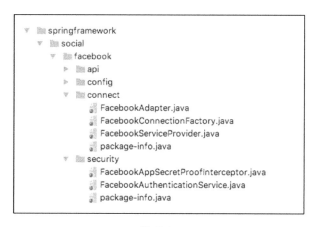

图 13-21

可以用下面的步骤实现一个简单的 OAuth 服务支持。

（1）创建源项目，如 spring-social-qq。

（2）添加服务提供商支持的 API，如 QQ。

（3）创建一个 ServiceProvider 模型，允许用户连接到服务提供商进行认证，以及获取授权的 API 实例，如 QQServiceProvider。

（4）创建一个 ApiAdapter，将服务提供商的 API 映射为统一的连接模型，如 QQAdapter。

（5）创建一个 ConnectionFactory 来整合上面这些部件，并提供一个简单的接口来创建连接，例如 QQConnectionFactory。

如果并非独立为某个 OAuth 服务做支持，而是希望在现有的项目中做集成开发，考虑到可能会有多个 OAuth 服务需要同时接入，那么可以参考图 13-22 所示的这种代码组织形式。

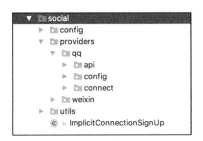

图 13-22

13.3.3 编码实现

在接入一个新的 OAuth 服务之前，首先需要了解如何与该服务提供商实施 API 交互，最直接的方式是查阅服务提供商的官方文档。OAuth 的整体交互流程并不复杂。首先，构建 URL 跳转到服务提供商的登录页面，在用户登录成功之后，携带 code 跳转回接入客户端指定的 URL；然后，Spring Social 的过滤器立刻获取该 code；接着，执行交换 accessToken，并获取 openId 的逻辑；最后，通过 accessToken 和 openId 无障碍地访问对应的资源服务器，获取信息。

1. 数据库准备

资源服务器提供的最基本的信息便是用户的个人资料，包括昵称、头像等。通常我们希望这些个人资料可以持久化，而不是每次用户登录时都去重新拉取。

一个用户的 accessToken、providerId（openId）、displayName、imageUrl 等信息组合在一起即可描述该用户的基本个人资料，Spring Social 称之为 Connections。对于这些 Connections 的持久化管理容器有一个标准的接口定义。

```
public interface ConnectionRepository {
    MultiValueMap<String, Connection<?>> findAllConnections();

    List<Connection<?>> findConnections(String var1);

    <A> List<Connection<A>> findConnections(Class<A> var1);

    MultiValueMap<String, Connection<?>>
findConnectionsToUsers(MultiValueMap<String, String> var1);

    Connection<?> getConnection(ConnectionKey var1);
```

```
    <A> Connection<A> getConnection(Class<A> var1, String var2);

    <A> Connection<A> getPrimaryConnection(Class<A> var1);

    <A> Connection<A> findPrimaryConnection(Class<A> var1);

    void addConnection(Connection<?> var1);

    void updateConnection(Connection<?> var1);

    void removeConnections(String var1);

    void removeConnection(ConnectionKey var1);
}
```

对于 ConnectionRepository，Spring Social 提供了两种实现方案：基于内存的 InMemoryConnectionRepository 和基于 JDBC 的 JdbcConnectionRepository。基于内存的实现方案缺陷较多，比如，Connections 在服务重启之后便会丢失，或者当长久运行时会占用过多内存空间等，因此更适用于演示。此处主要讲解基于 JDBC 的实现方案。

JdbcUsersConnectionRepository 默认实现了下面的数据库模型。

```
create table UserConnection (userId varchar(255) not null,
    providerId varchar(255) not null,
    providerUserId varchar(255),
    rank int not null,
    displayName varchar(255),
    profileUrl varchar(512),
    imageUrl varchar(512),
    accessToken varchar(512) not null,
    secret varchar(512),
    refreshToken varchar(512),
    expireTime bigint,
    primary key (userId, providerId, providerUserId));
create unique index UserConnectionRank on UserConnection(userId, providerId, rank);
```

这段代码位于 /org/springframework/social/connect/jdbc/JdbcUsersConnectionRepository.sql 中，尽管已经兼容了大多数的 SQL 数据库，但仍可能存在一些数据库特性不完全适用的情况，此时只需做出相应调整即可。

2. 创建 QQUserInfo

```java
// 对应 get_user_info 接口的数据形式
public class QQUserInfo {

    private String nickname;

    @JsonProperty("figureurl")
    private String figureUrl30;

    @JsonProperty("figureurl_1")
    private String figureUrl50;

    @JsonProperty("figureurl_2")
    private String figureUrl100;

    private String gender;

    @JsonProperty("figureurl_qq_1")
    private String qqFigureUrl40;

    @JsonProperty("figureurl_qq_2")
    private String qqFigureUrl100;

    // 携带 openId 备用
    private String openId;

    // 省略 setter 和 getter

}
```

3. 添加 RestTemplate 解析模板

在与 QQ 接口的交互上，响应类型都为 text/html 形式，RestTemplate 默认不支持该响应类型，所以应当自行添加。

主要有两类：一类是 text/html 转普通文本，另一类是 text/html 转 JSON 对象。

```java
public class TextHtmlHttpMessageConverter extends AbstractHttpMessageConverter
{

    public TextHtmlHttpMessageConverter() {
```

```java
        super(Charset.forName("UTF-8"), new MediaType[]{ MediaType.TEXT_HTML });
    }

    @Override
    protected boolean supports(Class clazz) {
        return String.class == clazz;
    }

    @Override
    protected Object readInternal(Class aClass, HttpInputMessage httpInputMessage) throws IOException, HttpMessageNotReadableException {
        Charset charset = this.getContentTypeCharset(httpInputMessage.getHeaders().getContentType());
        return StreamUtils.copyToString(httpInputMessage.getBody(), charset);
    }

    @Override
    protected void writeInternal(Object o, HttpOutputMessage httpOutputMessage) throws IOException, HttpMessageNotWritableException {
    }

    private Charset getContentTypeCharset(MediaType contentType) {
        return contentType != null && contentType.getCharset() != null?contentType.getCharset():this.getDefaultCharset();
    }

}
public class JacksonFromTextHtmlHttpMessageConverter extends MappingJackson2HttpMessageConverter {

    // 添加对 text/html 的支持
    public JacksonFromTextHtmlHttpMessageConverter() {
        List mediaTypes = new ArrayList();
        mediaTypes.add(MediaType.TEXT_HTML);
        setSupportedMediaTypes(mediaTypes);
    }

}
```

4. 创建 API 接口

```java
public interface QQ {

    QQUserInfo getUserInfo();
```

}

5. 实现并捆绑 API 接口

```java
// 继承 AbstractOAuth2ApiBinding，完成 API 接口与 Spring Social 的捆绑
public class QQImpl extends AbstractOAuth2ApiBinding implements QQ {

    // 获取 openId 的 API
    private static final String QQ_URL_GET_OPENID =
"https://graph.qq.com/oauth2.0/me?access_token={accessToken}";

    // 获取用户信息的 API
    private static final String QQ_URL_GET_USER_INFO =
"https://graph.qq.com/user/get_user_info?oauth_consumer_key={appId}&openid={openId}";

    private String appId;

    private String openId;

    // 构造函数在 accessToken 获取之后被调用，此处可以初始化 openId
    public QQImpl(String accessToken, String appId) {
        // access_token 作为查询参数被携带
        // 获取用户信息接口，无须声明携带 accessToken，所有资源交互接口都会自行携带
        // super(accessToken, TokenStrategy.ACCESS_TOKEN_PARAMETER);
        this.appId = appId;
        // 获取 openId 接口响应
        String result = getRestTemplate().getForObject(QQ_URL_GET_OPENID, String.class, accessToken);
        // 提取 openId
        this.openId = result.substring(result.lastIndexOf(":\"") + 2, result.indexOf("\"}"));
    }

    @Override
    public QQUserInfo getUserInfo() {
        QQUserInfo qqUserInfo = getRestTemplate().getForObject(QQ_URL_GET_USER_INFO, QQUserInfo.class, appId, openId);
        // 为用户信息类补充 openId
        if (qqUserInfo != null) {
            qqUserInfo.setOpenId(openId);
```

```
    }
    return qqUserInfo;
}


@Override
protected void configureRestTemplate(RestTemplate restTemplate) {
    // 为 API 交互的 RestTemplate 实例添加 text/html 转 JSON 对象的支持
    restTemplate.getMessageConverters().add(new
JacksonFromTextHtmlHttpMessageConverter());
}
}
```

6. 创建 QQServiceProvider

QQServiceProvider 包含 OAuth2Template 和 API 捆绑接口两个部件，所以在创建 QQServiceProvider 之前，应当实现 OAuth2Template。

```
public class QQOAuth2Template extends OAuth2Template {

    public QQOAuth2Template(String clientId, String clientSecret, String
authorizeUrl, String accessTokenUrl) {
        super(clientId, clientSecret, authorizeUrl, accessTokenUrl);
        // 指明获取 accessToken 时，需要携带 client_id 和 client_secret 两个参数（对应申
        // 请 QQ 互联时的 appId 和 appKey）
        setUseParametersForClientAuthentication(true);
    }

    @Override
    protected AccessGrant postForAccessGrant(String accessTokenUrl,
MultiValueMap<String, String> parameters) {
        // 获取 accessToken 接口的响应
        String responseStr = getRestTemplate().postForObject(accessTokenUrl,
parameters, String.class);

        // 从 API 文档中可以获知解析 accessToken 的方式
        String[] items = responseStr.split("&");
        //http://wiki.connect.qq.com/使用 authorization_code 获取 access_token
//access_token=FE04************************CCE2&expires_in=7776000&refresh_t
oken=88E4*********************BE14
        String accessToken = items[0].substring(items[0].lastIndexOf("=") + 1);
```

```
        Long expiresIn = new Long(items[1].substring(items[1].lastIndexOf("=")
+ 1));
        String refreshToken = items[2].substring(items[2].lastIndexOf("=") + 1);

        return new AccessGrant(accessToken, null, refreshToken, expiresIn);
    }

    /**
     *
     * 处理 text/html 类型的数据
     *
     * @return
     */

    @Override
    protected RestTemplate createRestTemplate() {
        RestTemplate restTemplate = super.createRestTemplate();
        restTemplate.getMessageConverters().add(new
TextHtmlHttpMessageConverter());
        return restTemplate;
    }
}
```

这里需要注意的地方有两点：一、RestTemplate 应当添加对 text/html 转普通文本的支持；二、accessToken 的解析需要自行实现。OAuth2Template 默认的实现方式并不适用于 QQ 的 OAuth 交互。

准备好 QQAuth2Template 之后，就可以创建 QQServiceProvider 了。

```
public class QQServiceProvider extends AbstractOAuth2ServiceProvider<QQ> {

    // 设置跳转 QQ 快捷登录页面的路径
    private static final String QQ_URL_AUTHORIZE =
"https://graph.qq.com/oauth2.0/authorize";

    // 获取 accessToken 接口
    private static final String QQ_URL_ACCESS_TOKEN =
"https://graph.qq.com/oauth2.0/token";

    private String appId;

    // 设置 OAuth2Template 的实现类
```

```
public QQServiceProvider(String appId, String appSecret) {
    super(new QQOAuth2Template(appId, appSecret, QQ_URL_AUTHORIZE,
QQ_URL_ACCESS_TOKEN));
    this.appId = appId;
}

// 声明 API 捆绑部件
@Override
public QQ getApi(String accessToken) {
    return new QQImpl(accessToken, appId);
}
}
```

7. 创建 ConnectionFactory

自行实现新的 OAuth 服务支持的过程其实是在 OAuth2ConnectionFactory 工厂类的模型下制造产品的过程，该工厂类描述了对服务提供商进行验证以及资源请求等所有细节。主要包含 ServiceProvider 和 ApiAdapter 两个部件。ServiceProvider 在前面的步骤中已经创建完毕。ApiAdapter 是什么呢？它实际上是一个可以令自定义 API 接口数据与 Spring Social 标准交互对象进行整合的适配器。例如，让不可预知的接口返回的用户信息填充为可预知的 Connection 对象等。

```
public class QQAdapter implements ApiAdapter<QQ> {

    @Override
    public boolean test(QQ qq) {
        return true;
    }

    // 通过 API 获取用户信息，并填充到 Connection 对象中
    @Override
    public void setConnectionValues(QQ qq, ConnectionValues connectionValues)
{
        QQUserInfo userInfo = qq.getUserInfo();
        connectionValues.setProviderUserId(userInfo.getOpenId());
        connectionValues.setDisplayName(userInfo.getNickname());
        connectionValues.setImageUrl(userInfo.getQqFigureUrl100());
        connectionValues.setProfileUrl(null);
    }

    @Override
```

```
    public UserProfile fetchUserProfile(QQ qq) {
        return null;
    }

    @Override
    public void updateStatus(QQ qq, String s) {

    }
}
```

接下来只需构建出 QQConnectionFactory 即可。

```
public class QQConnectionFactory extends OAuth2ConnectionFactory<QQ> {

    public QQConnectionFactory(String providerId, String appId, String appSecret) {
        super(providerId, new QQServiceProvider(appId, appSecret), new QQAdapter());
    }
}
```

8. 配置关于 QQ 的 Social 行为

关于 appId 和 appKey 等配置信息，建议抽象成一个配置类。

```
@Configuration
@ConfigurationProperties("spring.social.qq")
public class QQConfig {

    private String providerId = "qq";

    private String appId;

    private String appSecret;

    public String getAppId() {
        return appId;
    }

    public void setAppId(String appId) {
        this.appId = appId;
    }

    public String getAppSecret() {
```

```
        return appSecret;
    }

    public void setAppSecret(String appSecret) {
        this.appSecret = appSecret;
    }

    public String getProviderId() {
        return providerId;
    }

    public void setProviderId(String providerId) {
        this.providerId = providerId;
    }
}
```

创建 QQ 专用的连接工厂。

```
@Configuration
public class QQOAuth2Config extends SocialAutoConfigurerAdapter {

    @Autowired
    private QQConfig qqConfig;

    @Override
    protected ConnectionFactory<?> createConnectionFactory() {
        return new QQConnectionFactory(qqConfig.getProviderId(),
qqConfig.getAppId(), qqConfig.getAppSecret());
    }

}
```

9. 配置 Spring Social 的公共行为

前面的步骤完成之后，一个新的 OAuth 服务支持就结束了。但为了在项目中正式启用 Spring Social，还需要配置一些与项目相关的 Spring Social 公共行为。

配置 Spring Social 过滤器（SocialAuthenticationFilter）关注的一些 URL，包括访问哪些 URL 会触发 OAuth 逻辑，以及触发 OAuth 逻辑之后的一些跳转等，这部分内容定义在 SpringSocialConfigurer 中。默认情况下，SocialAuthenticationFilter 只关注/auth/{providerId}，当匹配正确时，会尝试获取 providerId，并执行一系列 OAuth 流程。如果 OAuth 认证后的用户在本地的用户数据库中无法找到，则认为这是一个需要注册的新用户，默认会跳转到/signup 这个路由下进行处理。

如果需要定制这些 URL，则可以通过继承 SpringSocialConfigurer 来实现。

```java
public class CommonSpringSocialConfigurer extends SpringSocialConfigurer {

    private String filterProcessesUrl;
    private String signUpUrl;


    public CommonSpringSocialConfigurer(String filterProcessesUrl, String signUpUrl) {
        this.filterProcessesUrl = filterProcessesUrl;
        this.signUpUrl = signUpUrl;
    }

    @Override
    protected <T> T postProcess(T object) {
        SocialAuthenticationFilter filter = (SocialAuthenticationFilter) super.postProcess(object);
        // 为 SocialAuthenticationFilter 设置自定义 URL
        filter.setFilterProcessesUrl(filterProcessesUrl);
        filter.setSignupUrl(signUpUrl);
        return (T) filter;
    }

}
```

与 Spring Security 类似，Spring Social 还提供了一个 @EnableSocial 注解，用于便捷导入 SocialConfiguration 类。只需继承 SocialConfigurerAdapter，即可实现在大部分行为默认的基础上定制部分配置的目的。

```java
@Configuration
@EnableSocial
public class SocialConfig extends SocialConfigurerAdapter {

    @Autowired
    private DataSource dataSource;

    @Autowired
    private ConnectionSignUp connectionSignUp;
```

```java
    // 声明自定义的SpringSocialConfigurer
    @Bean
    public SpringSocialConfigurer socialConfigurer() {
        return new CommonSpringSocialConfigurer("/oauth", "/register");
    }

    // 配置Spring Social 提供的OAuth登录工具
    @Bean
    public ProviderSignInUtils providerSignInUtils(ConnectionFactoryLocator factoryLocator) {
        return new ProviderSignInUtils(factoryLocator,
    getUsersConnectionRepository(factoryLocator));
    }

    @Override
    public UsersConnectionRepository
    getUsersConnectionRepository(ConnectionFactoryLocator
    connectionFactoryLocator) {
        // 声明一个基于JDBC 的用户连接管理容器
        JdbcUsersConnectionRepository repository = new
    JdbcUsersConnectionRepository(dataSource, connectionFactoryLocator,
    Encryptors.noOpText());
        // connections 隐式注册实例
        if (connectionSignUp != null) {
            repository.setConnectionSignUp(connectionSignUp);
        }
        return repository;
    }

}
```

ProviderSignInUtils 是一个用于处理 OAuth 登录逻辑的工具，主要通过 doPostSignUp 执行 Connection 持久化逻辑。

```java
public void doPostSignUp(String userId, RequestAttributes request) {
    ProviderSignInAttempt signInAttempt =
(ProviderSignInAttempt)this.sessionStrategy.getAttribute(request,
ProviderSignInAttempt.SESSION_ATTRIBUTE);
    if (signInAttempt != null) {
        signInAttempt.addConnection(userId, this.connectionFactoryLocator,
this.connectionRepository);
        this.sessionStrategy.removeAttribute(request,
ProviderSignInAttempt.SESSION_ATTRIBUTE);
```

        }
}

它并不会被主动调用,而是由客户端自行决定调用时机。例如,当 Spring Social 跳转到配置好的 signup 页面时,即可调用。

```
@GetMapping("/register")
public String socialRegister(ServletWebRequest request) {
    // 通过 request 对象获取 Connect
    Connection<?> connection =
providerSignInUtils.getConnectionFromSession(request);
    // 执行 Connection 持久化
    providerSignInUtils.doPostSignUp(connection.getKey().getProviderUserId(),
request);
    // 执行绑定账号、信息完善等其他逻辑
    ...

    // 跳转页面
    return "redirect:/";
}
```

当用户通过未在系统中使用过的第三方账号进行登录时,则可以"静默注册",实现 ConnectionSignUp 接口。

```
@Component
public class ImplicitConnectionSignUp implements ConnectionSignUp {

    @Override
    public String execute(Connection<?> connection) {
        // 用户名为 openId
        return connection.getKey().getProviderUserId();
    }
}
```

当用户再次登录时便不再跳转到 signup 页面,而是将 Connections 信息自动持久化到数据库中,并跳转回原访问页面。

13.4 与 Spring Security 整合

Spring Social 本身便是在 Spring Security 体系下开发的(它也可以在非 Spring Security 环境中工作),所以与 Spring Security 的整合极其简单。

```
@EnableWebSecurity(debug = true)
public class WebSecurityConfig extends WebSecurityConfigurerAdapter {

    @Autowired
    private SpringSocialConfigurer customSpringSocialConfigurer;

    ...

    @Override
    protected void configure(HttpSecurity http) throws Exception {
       http
          .authorizeRequests()
              ...
              // 开放Spring Social过滤器相关的URL访问权限
              .antMatchers("/register", "/oauth").permitAll()
              .anyRequest().authenticated()
              .and()
              .csrf().disable()
          .formLogin()
              ...
              .and()
          // 应用Spring Social配置
          .apply(customSpringSocialConfigurer);
    }

}
```

此外,由于Spring Security中有用户角色的概念,而默认情况下通过第三方登录的账号是没有用户角色的,所以要想真正进行整合,还应当让Spring Social的Connection与Spring Security的用户体系建立关系。

第1步,实现SocialUserDetailsService,允许Spring Social通过Connection的userId加载Spring Security体系下的用户信息。

```
// 基于Spring Security自定义数据库结构一节
@Service
public class MySocialUserDetailsService implements SocialUserDetailsService {

    @Autowired
    private UserService userService;

    @Override
    public SocialUserDetails loadUserByUserId(String s) throws
```

```
UsernameNotFoundException {
    User user = userService.findByUserName(s);
    if (user == null) {
        throw new UsernameNotFoundException("用户不存在");
    }
    // 设置权限

user.setAuthorities(AuthorityUtils.commaSeparatedStringToAuthorityList(user.getRoles()));
    return user;
    }
}
```

第 2 步，重新构建 ConnectionSignUp，在"静默注册"的同时生成用户记录，并赋予用户一个合适的角色。

```
@Component
public class ImplicitConnectionSignUp implements ConnectionSignUp {

    @Autowired
    UserService userService;

    @Override
    public String execute(Connection<?> connection) {
        // 隐式创建用户
        User user = new User(connection.getKey().getProviderUserId(),
"ROLE_USER");
        userService.createUser(user);
        return connection.getKey().getProviderUserId();
    }
}
```

13.5　Spring Social 源码分析

13.5.1　SocialAuthenticationFilter

SocialAuthenticationFilter 继承自 AbstractAuthenticationProcessingFilter，而 AbstractAuthenticationProcessingFilter 是 Spring Security 虚拟过滤器链上的专用过滤器实现。Spring Social 通过这种方式整合 Spring Security 的认证逻辑，默认的拦截地址是 /auth/{providerId}，可以通过配置 filterProcessesUrl 来更改前缀。

```java
private Authentication attemptAuthService(SocialAuthenticationService<?> 
authService, HttpServletRequest request, HttpServletResponse response) throws 
SocialAuthenticationRedirectException, AuthenticationException {
    // 通过OAuth2AuthenticationService获取SocialAuthenticationToken
    SocialAuthenticationToken token = authService.getAuthToken(request, 
response);
    if(token == null) {
        return null;
    } else {
        Assert.notNull(token.getConnection());
        Authentication auth = this.getAuthentication();
        if(auth != null && auth.isAuthenticated()) {
            this.addConnection(authService, request, token, auth);
            return null;
        } else {
            // 执行验证流程
            return this.doAuthentication(authService, request, token);
        }
    }
}

private Authentication doAuthentication(SocialAuthenticationService<?> 
authService, HttpServletRequest request, SocialAuthenticationToken token) {
    try {
        if(!authService.getConnectionCardinality().isAuthenticatePossible()) {
            return null;
        } else {

token.setDetails(this.authenticationDetailsSource.buildDetails(request));
            // 通过ProviderManger选举一个支持处理该token类型的
            //AuthenticationProvider，并进行验证
            Authentication success = 
this.getAuthenticationManager().authenticate(token);
            Assert.isInstanceOf(SocialUserDetails.class, 
success.getPrincipal(), "unexpected principle type");
            this.updateConnections(authService, token, success);
            return success;
        }
    } catch (BadCredentialsException var5) {
        if(this.signupUrl != null) {
            this.sessionStrategy.setAttribute(new ServletWebRequest(request), 
ProviderSignInAttempt.SESSION_ATTRIBUTE, new 
ProviderSignInAttempt(token.getConnection()));
```

```
        throw new 
SocialAuthenticationRedirectException(this.buildSignupUrl(request));
    } else {
        throw var5;
    }
  }
}
```

13.5.2　OAuth2AuthenticationService

OAuth2AuthenticationService 是 SocialAuthenticationService 接口的 OAuth 实现，主要围绕 OAuth2ConnectionFactory 来实现与 OAuth 授权服务器的交互。OAuth2ConnectionFactory 是 Spring Social 的连接工厂类，由两个关键部件组成：

◎　ApiAdapter

◎　OAuth2ServiceProvider

ApiAdapter，顾名思义，就是让 API 接口能够适配 OAuth 的标准验证流程。在 ApiAdapter 中，最为重要的一个方法是 setConnectionValues，该方法期望客户端用自行实现的 API 来填充 Spring Social 的 Connect 对象。

OAuth2ServiceProvider 包含了一个 OAuth2Template 类和一个 ApiBinding 接口，OAuth2Template 了客户端如何向授权服务器换取 accessToken，以及实现自动刷新 accessToken 的逻辑。

OAuth2Template 的实现是通用的，如果 OAuth 授权服务器提供的交互方式与 OAuth2Template 的不同，则可以通过继承 OAuth2Template 并覆写 postForAccessGrant 的方式来实现定制化的逻辑。

ApiBinding 主要用于描述资源提供商提供的 API 列表。

OAuth2AuthenticationService 的核心工作由 getAuthToken 方法完成。

```
// 获取 token
public SocialAuthenticationToken getAuthToken(HttpServletRequest request, 
HttpServletResponse response) throws SocialAuthenticationRedirectException {
    // 判断是否携带 code
    String code = request.getParameter("code");
    // 如果没有携带 code，则跳转到 OAuth2 服务提供商的授权页面
    if(!StringUtils.hasText(code)) {
```

```
        OAuth2Parameters params = new OAuth2Parameters();
        params.setRedirectUri(this.buildReturnToUrl(request));
        this.setScope(request, params);
        params.add("state", this.generateState(this.connectionFactory, request));
        this.addCustomParameters(params);
        throw new SocialAuthenticationRedirectException(this.getConnectionFactory().getOAuthOperations().buildAuthenticateUrl(params));
    } else if(StringUtils.hasText(code)) {
        try {
            String returnToUrl = this.buildReturnToUrl(request);
            // 调用 OAuth2Template 的 postForAccessGrant 来获取 accessToken
            AccessGrant accessGrant = this.getConnectionFactory().getOAuthOperations().exchangeForAccess(code, returnToUrl, (MultiValueMap)null);
            // 通过 ConnectionFactory 创建一个 OAuth2Connection 对象
            Connection<S> connection = this.getConnectionFactory().createConnection(accessGrant);
            return new SocialAuthenticationToken(connection, (Map)null);
        } catch (RestClientException var7) {
            this.logger.debug("failed to exchange for access", var7);
            return null;
        }
    } else {
        return null;
    }
}
```

13.5.3　OAuth2Connection

每个已登录用户都会维护一个 OAuth2Connection 对象，OAuth2Connection 通过构造函数初始化 accessToken 和 providerUserId 等信息。

```
public OAuth2Connection(String providerId, String providerUserId, String accessToken, String refreshToken, Long expireTime, OAuth2ServiceProvider<A> serviceProvider, ApiAdapter<A> apiAdapter) {
    super(apiAdapter);
    this.serviceProvider = serviceProvider;
    this.initAccessTokens(accessToken, refreshToken, expireTime);
    this.initApi();
    this.initApiProxy();
    this.initKey(providerId, providerUserId);
```

```java
}

// 实际上调用了 OAuth2ServiceProvider 的 getApi 方法
private void initApi() {
    this.api = this.serviceProvider.getApi(this.accessToken);
}

// 初始化 key
// key 包含了 providerId 和 providerUserId
protected void initKey(String providerId, String providerUserId) {
    // 如果首次获取不到 providerUserId
    if(providerUserId == null) {

        providerUserId = this.setValues().providerUserId;
    }

    this.key = new ConnectionKey(providerId, providerUserId);
}
private AbstractConnection<A>.ServiceProviderConnectionValuesImpl setValues()
{
    AbstractConnection<A>.ServiceProviderConnectionValuesImpl values = new
AbstractConnection.ServiceProviderConnectionValuesImpl();
    // 通过 ApiAdapter 的 setConnectionValues 获取 Connection 的值，接收一个定义好的
// API 接口
    // 可覆写该方法，实现自定义逻辑
    this.apiAdapter.setConnectionValues(this.getApi(), values);
    this.valuesInitialized = true;
    return values;
}
```

13.5.4　OAuth2Template

OAuth2Template 实现了 OAuth2Operations 接口，并定义了一系列与 accessToken 相关的逻辑。

```java
public AccessGrant exchangeForAccess(String authorizationCode, String
redirectUri, MultiValueMap<String, String> additionalParameters) {
    MultiValueMap<String, String> params = new LinkedMultiValueMap();
    // 如果设置了使用参数，则会携带 clientId 和 clientSecret
    // 并非所有 OAuth2 服务提供商都需要提供
    if(this.useParametersForClientAuthentication) {
```

```
        params.set("client_id", this.clientId);
        params.set("client_secret", this.clientSecret);
    }

    params.set("code", authorizationCode);
    params.set("redirect_uri", redirectUri);
    params.set("grant_type", "authorization_code");
    if(additionalParameters != null) {
        params.putAll(additionalParameters);
    }
    // 调用 postForAccessGrant,accessToken 接口默认返回的是 JSON 字符串
    // 可通过覆写来实现自定义逻辑
    return this.postForAccessGrant(this.accessTokenUrl, params);
}

// 默认的 accessToken 交换逻辑
protected AccessGrant postForAccessGrant(String accessTokenUrl,
MultiValueMap<String, String> parameters) {
    return
this.extractAccessGrant((Map)this.getRestTemplate().postForObject(accessToke
nUrl, parameters, Map.class, new Object[0]));
}

//该接口默认会返回一个 JSON 字符串,并解析为 map
private AccessGrant extractAccessGrant(Map<String, Object> result) {
    return this.createAccessGrant((String)result.get("access_token"),
(String)result.get("scope"), (String)result.get("refresh_token"),
this.getIntegerValue(result, "expires_in"), result);
}
```

13.5.5　SocialAuthenticationProvider

SocialAuthenticationProvider 实现了 AuthenticationProvider 接口,遵守的是 Spring Security 的认证逻辑。在 Spring Security 中,用户信息的交互是通过 UserDetailsService 实现的,Spring Social 定义了一个类似的接口 SocialUserDetailsService,我们需要实现 loadUserByUserId 方法,从而加载到该 OAuth 用户的权限等信息。

```
// 实现 authenticate 认证流程
public Authentication authenticate(Authentication authentication) throws
AuthenticationException {
    Assert.isInstanceOf(SocialAuthenticationToken.class, authentication,
"unsupported authentication type");
```

```java
    Assert.isTrue(!authentication.isAuthenticated(), "already authenticated");
    SocialAuthenticationToken authToken =
(SocialAuthenticationToken)authentication;
    String providerId = authToken.getProviderId();
    Connection<?> connection = authToken.getConnection();
    String userId = this.toUserId(connection);
    if(userId == null) {
        throw new BadCredentialsException("Unknown access token");
    } else {
        // 通过 SocialUserDetailsService 获取用户详细信息，包含授权角色、是否可用等属性
        UserDetails userDetails =
this.userDetailsService.loadUserByUserId(userId);
        if(userDetails == null) {
            throw new UsernameNotFoundException("Unknown connected account id");
        } else {
            return new SocialAuthenticationToken(connection, userDetails,
authToken.getProviderAccountData(), this.getAuthorities(providerId,
userDetails));
        }
    }
}

// 调用 UsersConnectionRepository，获取该 Connection 的 userId
protected String toUserId(Connection<?> connection) {
    List<String> userIds =
this.usersConnectionRepository.findUserIdsWithConnection(connection);
    return userIds.size() == 1?(String)userIds.iterator().next():null;
}
```

13.5.6　JdbcUsersConnectionRepository

　　JdbcUsersConnectionRepository 是 Spring Social 在用户通过 OAuth 认证后维护其信息的一种方式，包括用户信息的创建和查询。类似的实现还有 InMemoryUsersConnectionRepository。

```java
// 首次使用 OAuth 登录的用户会走注册逻辑，包括将记录存到 UserConnection 表
//以及调用 ConnectionSignUp 的 execute
// 可以在 execute 中实现用户的静默注册流程
public List<String> findUserIdsWithConnection(Connection<?> connection) {
    ConnectionKey key = connection.getKey();
    // 搜索 UserConnection 表
    List<String> localUserIds = this.jdbcTemplate.queryForList("select userId
from " + this.tablePrefix + "UserConnection where providerId = ? and
providerUserId = ?", String.class, new Object[]{key.getProviderId(),
```

```
key.getProviderUserId()});
    // 当表中不存在该记录时
    if(localUserIds.size() == 0 && this.connectionSignUp != null) {
        // 执行 ConnectSignUp 的 execute 方法，获取唯一用户 id
        String newUserId = this.connectionSignUp.execute(connection);
        if(newUserId != null) {
            // 如果可以正常获取用户 id，则把记录插入 UserConnection 表
this.createConnectionRepository(newUserId).addConnection(connection);
            return Arrays.asList(new String[]{newUserId});
        }
    }

    return localUserIds;
}
```

13.6　配置相关

1. @EnableSocial 注解

与 Spring Security 中的 @EnableWebSecurity 基本一致，都是通过注解引入配置类。

```
@Target({ElementType.TYPE})
@Retention(RetentionPolicy.RUNTIME)
@Documented
@Inherited
@Import({SocialConfiguration.class})
public @interface EnableSocial {
}
```

2. SocialConfiguration

SocialConfiguration 实际上是 Spring Social 的配置入口类，它与 @EnableSocial 注解配合，实现了对 Spring Social 的基本配置。

```
@Configuration
public class SocialConfiguration {
    private static boolean securityEnabled = isSocialSecurityAvailable();
    @Autowired
    private Environment environment;
    private List<SocialConfigurer> socialConfigurers;
```

```java
    public SocialConfiguration() {
    }

    // 获取所有SocialConfigurer配置类
    // 前面配置过的QQConnectionFactory也会被加载到此处
    @Autowired
    public void setSocialConfigurers(List<SocialConfigurer> socialConfigurers) {
        Assert.notNull(socialConfigurers, "At least one configuration class must implement SocialConfigurer (or subclass SocialConfigurerAdapter)");
        Assert.notEmpty(socialConfigurers, "At least one configuration class must implement SocialConfigurer (or subclass SocialConfigurerAdapter)");
        this.socialConfigurers = socialConfigurers;
    }

    // 携带ConnectionFactory的配置类，调用addConnectionFactories方法
    //把该工厂添加到ConnectionFactoryLocator
    @Bean
    public ConnectionFactoryLocator connectionFactoryLocator() {
        Iterator i$;
        SocialConfigurer socialConfigurer;
        if (securityEnabled) {
            SecurityEnabledConnectionFactoryConfigurer cfConfig = new SecurityEnabledConnectionFactoryConfigurer();
            i$ = this.socialConfigurers.iterator();

            while(i$.hasNext()) {
                socialConfigurer = (SocialConfigurer)i$.next();
                socialConfigurer.addConnectionFactories(cfConfig, this.environment);
            }

            return cfConfig.getConnectionFactoryLocator();
        } else {
            DefaultConnectionFactoryConfigurer cfConfig = new DefaultConnectionFactoryConfigurer();
            i$ = this.socialConfigurers.iterator();

            while(i$.hasNext()) {
                socialConfigurer = (SocialConfigurer)i$.next();
                socialConfigurer.addConnectionFactories(cfConfig, this.environment);
            }
```

```
            return cfConfig.getConnectionFactoryLocator();
        }
    }
    ...
}
```

在编写 QQOAuth2Config 时，覆写了 createConnectionFactory 方法，并返回一个 QQConnectionFactory。

```
@Configuration
public class QQOAuth2Config extends SocialAutoConfigurerAdapter {

    @Autowired
    private QQConfig qqConfig;

    @Override
    protected ConnectionFactory<?> createConnectionFactory() {
        return new QQConnectionFactory(qqConfig.getProviderId(),
qqConfig.getAppId(), qqConfig.getAppSecret());
    }

}
```

事实上，该方法是通过 SocialAutoConfigurerAdapter 的 addConnectionFactories 调用的。

```
public abstract class SocialAutoConfigurerAdapter extends
SocialConfigurerAdapter {
    public SocialAutoConfigurerAdapter() {
    }

    public void addConnectionFactories(ConnectionFactoryConfigurer configurer,
Environment environment) {
        configurer.addConnectionFactory(this.createConnectionFactory());
    }

    protected abstract ConnectionFactory<?> createConnectionFactory();
}
```

通过这种方式，最终所有声明过的连接工厂都会被 SocialConfiguration 自动获取。

3. SecurityEnabledConnectionFactoryConfigurer

SecurityEnabledConnectionFactoryConfigurer 是在 SocialConfiguration 中被调用的，它首先把 ConnectionFactory 注入 OAuth2AuthenticationService 中，接着又把 OAuth2AuthenticationService 添加到 SocialAuthenticationServiceRegistry 中，并返回给 SocialConfiguration。SocialConfiguration 将其配置为 bean，以便后续使用。

```
class SecurityEnabledConnectionFactoryConfigurer implements
ConnectionFactoryConfigurer {
    private SocialAuthenticationServiceRegistry registry = new
SocialAuthenticationServiceRegistry();

    public SecurityEnabledConnectionFactoryConfigurer() {
    }

    public void addConnectionFactory(ConnectionFactory<?> connectionFactory) {
this.registry.addAuthenticationService(this.wrapAsSocialAuthenticationServic
e(connectionFactory));
    }

    public ConnectionFactoryRegistry getConnectionFactoryLocator() {
        return this.registry;
    }

    // 使用 ConnectionFactory 构建 OAuth2AuthenticationService
    private <A> SocialAuthenticationService<A>
wrapAsSocialAuthenticationService(ConnectionFactory<A> cf) {
        if (cf instanceof OAuth1ConnectionFactory) {
            return new OAuth1AuthenticationService((OAuth1ConnectionFactory)cf);
        } else if (cf instanceof OAuth2ConnectionFactory) {
            OAuth2AuthenticationService<A> authService = new
OAuth2AuthenticationService((OAuth2ConnectionFactory)cf);
authService.setDefaultScope(((OAuth2ConnectionFactory)cf).getScope());
            return authService;
        } else {
            throw new IllegalArgumentException("The connection factory must be one
of OAuth1ConnectionFactory or OAuth2ConnectionFactory");
        }
    }
}
```

第 4 部分

第14章
用Spring Security OAuth实现OAuth对接

Spring Security OAuth 是一个专注于 OAuth 认证的框架，它完整覆盖了客户端、资源服务和认证服务三个模块。这三个模块分别在 Spring Security 5.0、5.1 和 5.3（目前还在计划开发中）三个版本中被集成，原有的独立项目则进入维护状态。

Spring Security 5.0 中集成了 OAuth 的客户端模块，该模块包含以下三个子模块。

（1）spring-security-oauth2-core：OAuth 授权框架和 OIDC 的核心数据结构及接口，被 Client、Resource Server 和 Authorization Server 所依赖。

（2）spring-security-oauth2-jose：支持 JOSE 协议组，具体包括以下内容。

- ◎ JSON Web Token (JWT)；
- ◎ JSON Web Signature (JWS)；
- ◎ JSON Web Encryption (JWE)；
- ◎ JSON Web Key (JWK)。

（3）spring-security-oauth2-client：是 Spring Security 支持 OAuth 和 OIDC 的客户端功能实现包。

下面将快速体验 Spring Security 的 OAuth 功能。考虑接入成本等问题，推荐使用 GitHub 作为 OAuth 服务提供商。因为 GitHub 提供的认证服务符合标准的 OAuth，并且 GitHub 账号申请较为方便，有利于聚焦 Spring Security 的 OAuth 客户端接入流程。

14.1　实现 GitHub 快捷登录

1. 新建工程

首先，新建 Spring Boot 2.0 工程，pom 包依赖如下。

```xml
<dependency>
        <groupId>org.springframework.boot</groupId>
        <artifactId>spring-boot-starter-web</artifactId>
</dependency>
<dependency>
        <groupId>org.springframework.boot</groupId>
        <artifactId>spring-boot-starter-security</artifactId>
</dependency>
<dependency>
        <groupId>org.springframework.security</groupId>
        <artifactId>spring-security-config</artifactId>
</dependency>
<dependency>
        <groupId>org.springframework.security</groupId>
        <artifactId>spring-security-oauth2-client</artifactId>
</dependency>
<dependency>
        <groupId>org.springframework.security</groupId>
        <artifactId>spring-security-oauth2-jose</artifactId>
</dependency>
<dependency>
        <groupId>org.springframework.boot</groupId>
        <artifactId>spring-boot-starter-test</artifactId>
        <scope>test</scope>
</dependency>
```

2. 注册 OAuth 应用

在 GitHub 官网上注册一个新的 OAuth 应用，地址是 https://github.com/settings/applications/new，打开页面如图 14-1 所示。

Application name：应用名称，必填项。

Homepage URL：主页 URL，必填项。在本地开发时，将其设置为 http://localhost:8080 即可。

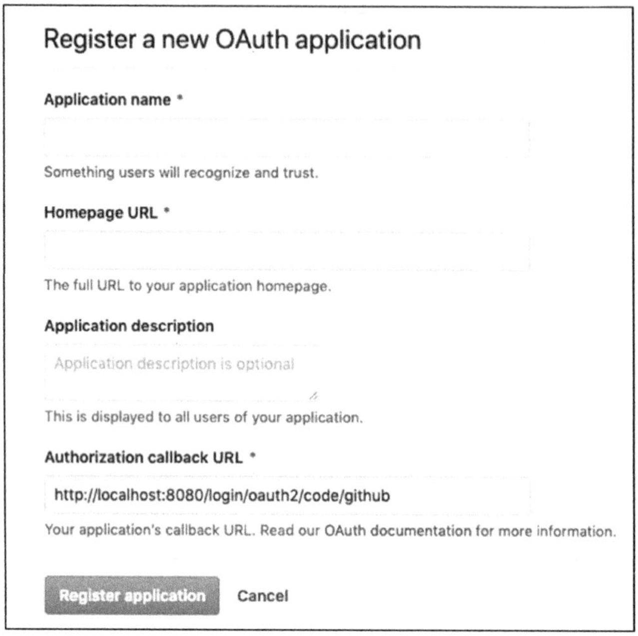

图 14-1

Application description：应用的说明，选填项，置空即可。

Authorization callback URL：OAuth 认证的重定向地址，必填项，本地开发环节可设置为 http://localhost:8080/login/oauth2/code/github。

单击"Register application"按钮，即可注册得到 clientId 和 clientSecret。

当用户通过用户代理（浏览器）成功登录 GitHub，并且用户在批准页（Approva Page）授权允许注册的客户端应用访问自己的用户数据后，GitHub 会将授权码（Code）通过重定向的方式传递给客户端应用。

Spring Security OAuth 默认的重定向模板是 {baseUrl}/login/oauth2/code/{registrationId}，registrationId 是 ClientRegistration 的唯一 ID，通常以接入的 OAuth 服务提供商的简称来命名即可，所以此处设置为 github。

3. 配置 application.yml

前面在工程的 pom 文件中引入了相关依赖包，并且在 GitHub 上成功注册了一个 OAuth 客户端应用，接下来需要在配置文件 application.yml 中增加相应配置。

```yaml
spring:
  security:
    oauth2:
      client:
        registration:          (1)
          github:              (2)
            client-id: github-client-id
            client-secret: github-client-secret
```

说明：

（1）spring.security.oauth2.client.registration 是 OAuth 客户端所有属性的基础前缀。

（2）registration 下面的 github 是 ClientRegistration 的唯一 ID。

另外，client-id 和 client-secret 需要替换为前面在 GitHub 上注册得到的 clientId 和 clientSecret。

4．新建 Controller

```java
@RestController
public class SimpleController {

    @GetMapping("/hello")
    public String hello(Principal principal) {
        return "hello, " + principal.getName();
    }

}
```

参数中的 principal 对象由 Spring 框架自动注入，表示当前登录的用户。

5．效果演示

（1）启动新建的 OAuth 工程。

（2）用浏览器访问 http://localhost:8080/hello 时，会重定向到默认生成的登录页，单击 "GitHub" 按钮，即可跳转到 GitHub 登录页，如图 14-2 所示。

当认证成功时将会跳转到确认授权页面，如图 14-3 所示。

图 14-2　　　　　　　　　　　　　图 14-3

单击"Authorize andyzhaozhao"按钮，以允许 OAuth 客户端应用访问 GitHub 的用户数据。此时 OAuth 客户端应用会调用用户数据接口（the UserInfo Endpoint），创建认证对象。浏览器最终将自动重定向到原访问地址 http://localhost:8080/hello，并打印字符串"hello，×××"。

14.2　用 Spring Security OAuth 实现 QQ 快捷登录

前面使用最简配置实现了 OAuth 客户端接入 GitHub 登录的功能，得益于 Spring Security 对 OAuth 标准认证流程的封装，最简配置客户端也可以很方便地接入 Google 和 Facebook 等实现相对标准的 OAuth 服务提供商。而对于 QQ 登录等 OAuth 流程来说，则可以在此基础上进行额外的适配工作，Spring Security 良好的 OAuth 扩展性同样为适配提供了足够的支持。

14.2.1　OAuth 功能扩展流程

不同 OAuth 服务提供商提供的授权流程在细节上略有不同，主要体现在与授权服务器交互过程中的传参、返回值解析，以及从资源服务器获取资源等几个方面。核心步骤大体相同，都是 OAuth 标准制定的形式。

◎ 获取 code。
◎ 使用 code 交换 access_token。
◎ 携带 access_token 请求被保护的用户信息和其他资源。

针对这三个核心步骤，Spring Security 提供了相应的扩展接口和配置方法。

（1）支持自定义重定向端点（Redirection Endpoint）。OAuth 服务器通过重定向到此端点的方式将 code 传递给 OAuth 客户端。

（2）支持自定义 OAuth2AccessTokenResponseClient，OAuth2AccessTokenResponseClient 实现了以 code 交换 access_token 的具体逻辑。

（3）支持自定义用户信息端点（UserInfo Endpoint）。常用自定义方式如下。

◎ 自定义 OAuth2User。不同 OAuth 服务提供商的用户属性不同，可以针对不同的 OAuth 服务提供商做适配。
◎ 自定义 OAuth2UserService。OAuth2UserService 负责请求用户信息（OAuth2User）。标准 OAuth 可以直接携带 access_token 请求用户信息，而 QQ 则需要先获取 OpenId，再使用 OpenId 获取用户信息。

一般来说，在对接社交账号的登录功能时，不会局限于单个 OAuth 服务提供商，而是同时提供多种流行的社交平台供用户选择，所以需要有多套 OAuth 方案并存的准备。为了避免项目中不同 OAuth 对接代码混乱的情况，推荐使用图 14-4 所示的这种组织形式。

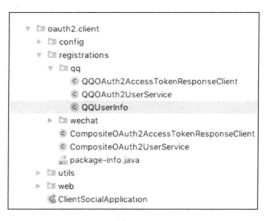

图 14-4

另外，在使用 Spring Social 对接 QQ 登录时，基本的准备工作都有提及，读者可以直接参考该部分内容。

14.2.2 编码实现

相对于标准的 OAuth 授权码模式，QQ 提供的 API 在交互上较为混乱，其响应类型为 text/html，响应内容有普通文本、JSON 字符串等多种类型。另外，QQ 提供的 API 需要先获取 OpenId，再使用 OpenId 结合 appId 与 access_token 的方式来获取用户信息，而不是直接使用 access_token，这些都是在自定义实现时需要注意的内容。

1. 新建项目

首先，新建 Spring Boot 2.0 工程，命名为 client-social，引入 spring-boot-starter-web 和 spring-boot-starter-security 两个依赖包。

```
<dependency>
        <groupId>org.springframework.boot</groupId>
        <artifactId>spring-boot-starter-web</artifactId>
</dependency>
<dependency>
        <groupId>org.springframework.boot</groupId>
        <artifactId>spring-boot-starter-security</artifactId>
</dependency>
```

引入 Spring Security 为 OAuth 提供支持的专用依赖包。

```
<dependency>
        <groupId>org.springframework.security</groupId>
        <artifactId>spring-security-config</artifactId>
</dependency>
<dependency>
        <groupId>org.springframework.security</groupId>
        <artifactId>spring-security-oauth2-client</artifactId>
</dependency>
<dependency>
        <groupId>org.springframework.security</groupId>
        <artifactId>spring-security-oauth2-jose</artifactId>
</dependency>
```

2. 自定义 QQUserInfo 实现 OAuth2User 接口

因为无法使用默认的 DefaultOAuth2User 来表示 QQ 用户信息，所以需要提供一个自定义的 QQUserInfo 类并实现 OAuth2User 接口。

```java
public class QQUserInfo implements OAuth2User {

    // 统一赋予 USER 角色
    private List<GrantedAuthority> authorities =
            AuthorityUtils.createAuthorityList("ROLE_USER");
    private Map<String, Object> attributes;

    private String nickname;
    @JsonProperty("figureurl")
    private String figureUrl30;
    @JsonProperty("figureurl_1")
    private String figureUrl50;
    @JsonProperty("figureurl_2")
    private String figureUrl100;
    @JsonProperty("figureurl_qq_1")
    private String qqFigureUrl40;
    @JsonProperty("figureurl_qq_2")
    private String qqFigureUrl100;
    private String gender;
    // 携带 openId 备用
    private String openId;

    @Override
    public Collection<? extends GrantedAuthority> getAuthorities() {
        return this.authorities;
    }

    @Override
    public Map<String, Object> getAttributes() {
        if (this.attributes == null) {
            this.attributes = new HashMap<>();
            this.attributes.put("nickname", this.getNickname());
            this.attributes.put("figureUrl30", this.getFigureUrl30());
            this.attributes.put("figureUrl50", this.getFigureUrl50());
            this.attributes.put("figureUrl100", this.getFigureUrl100());
            this.attributes.put("qqFigureUrl40", this.getQqFigureUrl40());
            this.attributes.put("qqFigureUrl100", this.getQqFigureUrl100());
            this.attributes.put("gender", this.getGender());
```

```
            this.attributes.put("openId", this.getOpenId());
        }
        return attributes;
    }

    @Override
    public String getName() {
        return this.nickname;
    }

    // 省略 setter 和 getter
}
```

3. 添加 RestTemplate 解析模板

```
public class JacksonFromTextHtmlHttpMessageConverter extends
MappingJackson2HttpMessageConverter {

    // 添加对 text/html 的支持
    public JacksonFromTextHtmlHttpMessageConverter() {
        List mediaTypes = new ArrayList();
        mediaTypes.add(MediaType.TEXT_HTML);
        setSupportedMediaTypes(mediaTypes);
    }

}
public class TextHtmlHttpMessageConverter extends AbstractHttpMessageConverter
{

    public TextHtmlHttpMessageConverter() {
        super(Charset.forName("UTF-8"), new MediaType[]{MediaType.TEXT_HTML});
    }

    @Override
    protected boolean supports(Class clazz) {
        return String.class == clazz;
    }

    @Override
    protected Object readInternal(Class aClass, HttpInputMessage
httpInputMessage) throws IOException, HttpMessageNotReadableException {
        Charset charset =
this.getContentTypeCharset(httpInputMessage.getHeaders().getContentType());
```

```
        return StreamUtils.copyToString(httpInputMessage.getBody(), charset);
    }

    @Override
    protected void writeInternal(Object o, HttpOutputMessage httpOutputMessage)
throws IOException, HttpMessageNotWritableException {
    }

    private Charset getContentTypeCharset(MediaType contentType) {
        return contentType != null && contentType.getCharset() != null ?
contentType.getCharset() : this.getDefaultCharset();
    }
}
```

4. 自定义实现 OAuth2AccessTokenResponseClient 接口

OAuth2AccessTokenResponseClient 实现了以 code 交换 access_token 的具体逻辑。默认提供的实现类 NimbusAuthorizationCodeTokenResponseClients 可以实现标准的 OAuth 交换 access_token 的具体逻辑，但 QQ 提供的方式并不标准，所以需要自定义实现 OAuth2AccessTokenResponseClient 接口。

```
public class QQOAuth2AccessTokenResponseClient implements
OAuth2AccessTokenResponseClient<OAuth2AuthorizationCodeGrantRequest> {

    private RestTemplate restTemplate;

    private RestTemplate getRestTemplate() {
        if (restTemplate == null) {
            restTemplate = new RestTemplate();
            restTemplate.getMessageConverters().add(new
TextHtmlHttpMessageConverter());
        }

        return restTemplate;
    }

    @Override
    public OAuth2AccessTokenResponse
getTokenResponse(OAuth2AuthorizationCodeGrantRequest
authorizationGrantRequest)
            throws OAuth2AuthenticationException {
        ClientRegistration clientRegistration =
```

```
authorizationGrantRequest.getClientRegistration();
        OAuth2AuthorizationExchange oAuth2AuthorizationExchange =
authorizationGrantRequest.getAuthorizationExchange();

        // 根据 API 文档获取请求 access_token 参数
        MultiValueMap<String, String> params = new LinkedMultiValueMap();
        params.set("client_id", clientRegistration.getClientId());
        params.set("client_secret", clientRegistration.getClientSecret());
        params.set("code",
oAuth2AuthorizationExchange.getAuthorizationResponse().getCode());
        params.set("redirect_uri",
oAuth2AuthorizationExchange.getAuthorizationRequest().getRedirectUri());
        params.set("grant_type", "authorization_code");
        String tmpTokenResponse =
getRestTemplate().postForObject(clientRegistration.getProviderDetails().getT
okenUri(), params, String.class);

        // 从 API 文档中可以获知解析 accessToken 的方式
        String[] items = tmpTokenResponse.split("&");
        //http://wiki.connect.qq.com/使用 authorization_code 获取
//access_token
        //access_token=FE04************************CCE2&expires_in=7776
//000&refresh_token=88E4************************BE14
        String accessToken = items[0].substring(items[0].lastIndexOf("=") + 1);
        Long expiresIn = new Long(items[1].substring(items[1].lastIndexOf("=")
+ 1));

        Set<String> scopes = new
LinkedHashSet<>(oAuth2AuthorizationExchange.getAuthorizationRequest().getSco
pes());
        Map<String, Object> additionalParameters = new LinkedHashMap<>();
        OAuth2AccessToken.TokenType accessTokenType =
OAuth2AccessToken.TokenType.BEARER;

        return OAuth2AccessTokenResponse.withToken(accessToken)
                .tokenType(accessTokenType)
                .expiresIn(expiresIn)
                .scopes(scopes)
                .additionalParameters(additionalParameters)
                .build();
    }
}
```

主要使用 RestTemplate 请求获取 access_token，并对返回的结果执行自定义解析，最后构建成 OAuth2AccessTokenResponse 对象返回即可。

5. 实现 OAuth2UserService 接口

OAuth2UserService 负责请求用户信息（OAuth2User）。标准的 OAuth 可以直接携带 access_token 请求用户信息，但 QQ 需要获取到 OpenId 才能使用。

```
public class QQOAuth2UserService implements
OAuth2UserService<OAuth2UserRequest, OAuth2User> {
    // 获取用户信息的 API
    private static final String QQ_URL_GET_USER_INFO =
"https://graph.qq.com/user/get_user_info?oauth_consumer_key={appId}&openid={openId}&access_token={access_token}";

    private RestTemplate restTemplate;

    private RestTemplate getRestTemplate() {
        if (restTemplate == null) {
            restTemplate = new RestTemplate();
            //通过 Jackson JSON processing library 直接将返回值绑定到对象
            restTemplate.getMessageConverters().add(new
JacksonFromTextHtmlHttpMessageConverter());
        }

        return restTemplate;
    }

    @Override
    public OAuth2User loadUser(OAuth2UserRequest userRequest) throws
OAuth2AuthenticationException {
        // 第一步：获取 openId 接口响应
        String accessToken = userRequest.getAccessToken().getTokenValue();
        String openIdUrl =
userRequest.getClientRegistration().getProviderDetails().getUserInfoEndpoint().getUri() + "?access_token={accessToken}";
        String result = getRestTemplate().getForObject(openIdUrl, String.class,
accessToken);
        // 提取 openId
        String openId = result.substring(result.lastIndexOf(":\"") + 2,
result.indexOf("\"}"));

        // 第二步：获取用户信息
        String appId = userRequest.getClientRegistration().getClientId();
```

```
        QQUserInfo qqUserInfo = 
getRestTemplate().getForObject(QQ_URL_GET_USER_INFO, QQUserInfo.class, appId, 
openId, accessToken);
        // 为用户信息类补充 openId
        if (qqUserInfo != null) {
            qqUserInfo.setOpenId(openId);
        }
        return qqUserInfo;
    }
}
```

首先获取 OpenId,然后通过 OpenId 等参数获取用户信息,最终组装成 QQUserInfo 对象。

6. 多个 OAuth 服务提供商并存

前面我们通过自定义实现 QQOAuth2AccessTokenResponseClient 和 QQOAuth2UserService 来支持 QQ 登录,但如果直接使用它们分别代替默认的 NimbusAuthorizationCodeTokenResponseClient 和 DefaultOAuth2UserService,将会导致 GitHub 等标准 OAuth 服务无法正常使用。为了让多个 OAuth 服务可以并存,建议使用组合模式。

(1)提供 CompositeOAuth2AccessTokenResponseClient。

```java
/**
 * OAuth2AccessTokenResponseClient 的组合类,使用了 Composite Pattern (组合模式)。
 * 除支持 Google、Okta、GitHub 和 Facebook 外,还支持 QQ、微信等多种认证服务,
 * 可根据 registrationId 选择相应的 OAuth2AccessTokenResponseClient
 */
public class CompositeOAuth2AccessTokenResponseClient implements 
OAuth2AccessTokenResponseClient<OAuth2AuthorizationCodeGrantRequest> {

    private Map<String, OAuth2AccessTokenResponseClient> clients;

    private static final String DefaultClientKey = "default_key";

    public CompositeOAuth2AccessTokenResponseClient() {
        this.clients = new HashMap();
        // spring-security-oauth2-client 默认的 OAuth2AccessTokenResponse
        //Client 实现类是 NimbusAuthorizationCodeTokenResponseClient
        // 将其预置到组合类 CompositeOAuth2AccessTokenResponseClient 中,
        //使其默认支持 Google、Okta、GitHub 和 Facebook
        this.clients.put(DefaultClientKey, new 
NimbusAuthorizationCodeTokenResponseClient());
```

```
    }

    @Override
    public OAuth2AccessTokenResponse
getTokenResponse(OAuth2AuthorizationCodeGrantRequest
authorizationGrantRequest)
            throws OAuth2AuthenticationException {
        ClientRegistration clientRegistration =
authorizationGrantRequest.getClientRegistration();

        OAuth2AccessTokenResponseClient client =
clients.get(clientRegistration.getRegistrationId());
        if (client == null) {
            client = clients.get(DefaultClientKey);
        }

        return client.getTokenResponse(authorizationGrantRequest);
    }

    public Map<String, OAuth2AccessTokenResponseClient>
getOAuth2AccessTokenResponseClients() {
        return clients;
    }
}
```

（2） 提供 CompositeOAuth2UserService。

```
public class CompositeOAuth2UserService implements
OAuth2UserService<OAuth2UserRequest, OAuth2User> {
    private Map<String, OAuth2UserService> userServices;

    private static final String DefaultUserServiceKey = "default_key";

    public CompositeOAuth2UserService() {
        this.userServices = new HashMap();
        // DefaultOAuth2UserService 是默认处理 OAuth 获取用户逻辑的
        // OAuth2UserService 实现类
        // 将其预置到组合类 CompositeOAuth2UserService 中，从而默认支持 Google、Okta、
        // GitHub 和 Facebook
        this.userServices.put(DefaultUserServiceKey, new
DefaultOAuth2UserService());
    }
```

```java
    @Override
    public OAuth2User loadUser(OAuth2UserRequest userRequest) throws OAuth2AuthenticationException {
        ClientRegistration clientRegistration = userRequest.getClientRegistration();

        OAuth2UserService service = userServices.get(clientRegistration.getRegistrationId());
        if (service == null) {
            service = userServices.get(DefaultUserServiceKey);
        }

        return service.loadUser(userRequest);
    }

    public Map<String, OAuth2UserService> getUserServices() {
        return userServices;
    }
}
```

7. 配置 Spring Security

从 Spring Security 5.0 开始，在 HttpSecurity 中提供了用于配置 OAuth 客户端的策略 OAuth2Login()方法。

```
http.OAuth2Login()
```

完整配置如下。

```java
@EnableWebSecurity(debug = true)
public class SecurityConfig extends WebSecurityConfigurerAdapter {

    public static final String QQRegistrationId = "qq";
    public static final String WeChatRegistrationId = "wechat";

    public static final String LoginPagePath = "/login/oauth2";

    @Override
    public void configure(HttpSecurity http) throws Exception {
        http.authorizeRequests()
                .antMatchers(LoginPagePath).permitAll()
                .anyRequest()
                .authenticated();
        http.oauth2Login()
```

```java
            // 使用 CompositeOAuth2AccessTokenResponseClient
            .tokenEndpoint().accessTokenResponseClient(this.accessTokenRes
                                                        ponseClient())
            .and()
            .userInfoEndpoint()
            .customUserType(QQUserInfo.class, QQRegistrationId)
            // 使用 CompositeOAuth2UserService
            .userService(oauth2UserService())
            // 可选，要保证与 redirect-uri-template 匹配
            .and()
        .redirectionEndpoint().baseUri("/register/social/*");

    //自定义登录页
    http.oauth2Login().loginPage(LoginPagePath);
}

private
OAuth2AccessTokenResponseClient<OAuth2AuthorizationCodeGrantRequest>
accessTokenResponseClient() {
    CompositeOAuth2AccessTokenResponseClient client = new
CompositeOAuth2AccessTokenResponseClient();
    // 加入 QQ 自定义 QQOAuth2AccessTokenResponseClient
    client.getOAuth2AccessTokenResponseClients().put(QQRegistrationId, new
QQOAuth2AccessTokenResponseClient());
    return client;
}

private OAuth2UserService<OAuth2UserRequest, OAuth2User>
oauth2UserService() {
    CompositeOAuth2UserService service = new CompositeOAuth2UserService();
    // 加入 QQ 自定义 QQOAuth2UserService
    service.getUserServices().put(QQRegistrationId, new
QQOAuth2UserService());
    return service;
}
}
```

其中，关于重定向端点（redirectionEndpoint）的配置是可选的，需要注意的是，当多个 OAuth 服务提供商并存时，一定要保证 baseUri、redirect-uri-template 和 OAuth 注册的重定向地址三者相互匹配。

8. 工程配置文件

下面的配置同时支持 GitHub 和 QQ 登录。

```yaml
server:
  port: 8080

logging:
  level:
    root: INFO
    org.springframework.web: INFO
    org.springframework.security: DEBUG
    org.springframework.boot.autoconfigure: DEBUG

spring:
  security:
    oauth2:
      client:
        registration:
          github:
            client-id: {custom}
            client-secret: {custom}
            redirect-uri-template:
"{baseUrl}/register/social/{registrationId}"
          qq:
            client-id: {custom appId}
            client-secret: {custom appKey}
            provider: qq
            client-name: QQ登录
            authorization-grant-type: authorization_code
            client-authentication-method: post
            scope: get_user_info,list_album,upload_pic,do_like
            redirect-uri-template:
"{baseUrl}/register/social/{registrationId}"
        provider:
          qq:
            authorization-uri: https://graph.qq.com/oauth2.0/authorize
            token-uri: https://graph.qq.com/oauth2.0/token
            # 配置为 QQ 获取 OpenId 的 URL
            user-info-uri: https://graph.qq.com/oauth2.0/me
            user-name-attribute: "nickname"
```

14.2.3 自定义 login.html 和 index.html

Spring Security 的 OAuth 功能通过 DefaultLoginPageGeneratingFilter 生成了一个默认的登录页，除此之外，也可以自定义登录页。

用工程的 pom 文件引入 thymeleaf 模板。

```xml
<!--页面模板-->
<dependency>
        <groupId>org.springframework.boot</groupId>
        <artifactId>spring-boot-starter-thymeleaf</artifactId>
</dependency>
<dependency>
        <groupId>org.thymeleaf.extras</groupId>
        <artifactId>thymeleaf-extras-springsecurity4</artifactId>
</dependency>
```

index.html 为主页面，主要用于展示信息。

```html
<!DOCTYPE html>
<html xmlns="http://www.w3.org/1999/xhtml"
xmlns:th="http://www.thymeleaf.org"
    xmlns:sec="http://www.thymeleaf.org/thymeleaf-extras-springsecurity4">
   <head>
        <title>Spring Security - OAuth 2.0 Login</title>
        <meta charset="utf-8"/>
   </head>
   <body>
        <div style="float: right" th:fragment="logout" sec:authorize="isAuthenticated()">
            <div style="float:left">
                <span style="font-weight:bold">用户：</span><span sec:authentication="name"></span>
            </div>
            <div style="float:none"> </div>
            <div style="float:right">
                <form action="#" th:action="@{/logout}" method="post">
                    <input type="submit" value="注销"/>
                </form>
            </div>
        </div>
        <h1>使用 Spring Security OAuth 2.0 登录</h1>
        <div>
```

```
            恭喜您通过"<span style="font-weight:bold"
th:text="${clientName}"></span>"
           登录成功
       </div>
    </body>
</html>
```

login.html 是我们自定义的登录页。

```
<!DOCTYPE html>
<html xmlns="http://www.w3.org/1999/xhtml">
    <head>
        <title>Spring Security - OAuth 2.0 Login</title>
        <meta charset="utf-8"/>
    </head>
    <body>
        <h1>自定义 OAuth2 登录页</h1>
        <div>
            <a href="/oauth2/authorization/github">GitHub 登录</a>
            <a href="/oauth2/authorization/qq">QQ 登录</a>
        </div>
    </body>
</html>
```

14.2.4　自定义 Controller 映射

```
@Controller
public class MainController {

    @Autowired
    private OAuth2AuthorizedClientService authorizedClientService;

    @GetMapping("/")
    public String index(Model model, OAuth2AuthenticationToken authentication) {
        OAuth2AuthorizedClient authorizedClient = this.getAuthorizedClient(authentication);
        model.addAttribute("userName", authentication.getName());
        model.addAttribute("clientName", authorizedClient.getClientRegistration().getClientName());
        return "index";
    }
```

```
@GetMapping("/login/oauth2")
public String login() {
    return "login";
}

private OAuth2AuthorizedClient
getAuthorizedClient(OAuth2AuthenticationToken authentication) {
    return this.authorizedClientService.loadAuthorizedClient(
            authentication.getAuthorizedClientRegistrationId(),
authentication.getName());
   }
}
```

OAuth2AuthenticationToken 可以获取当前用户信息，由 Spring Security 自动注入。OAuth2AuthorizedClientService 对象可以用来获取当前已经认证成功的 OAuth 客户端信息。

14.2.5　启用自定义登录页

```
http.oauth2Login()
        .loginPage("/login/oauth2")
```

（1）运行 client-social 工程。

（2）在浏览器地址栏输入 http://{ip|host}:{port}/，得到如图 14-5 所示页面。

图 14-5

注意，在 QQ 互联平台注册网站回调域时，填写的回调地址应配置在本地 hosts 中，并用该域名进行访问测试，直接通过 localhost 测试是无法成功的。

（3）单击 GitHub 登录按钮或者 QQ 登录按钮，按照提示进行登录操作。在登录成功后会跳转到主页，如图 14-6 所示。

![图 14-6](使用Spring Security OAuth 2.0 登录 用户: 7730080 注销 恭喜您通过"GitHub"登录成功)

图 14-6

14.3　OAuth Client 功能核心源码分析

前面对 Spring Security OAuth Client 功能进行了扩展开发，从而支持 QQ 登录，同时实现了多个 OAuth 服务提供商共存的效果。要想了解这些扩展方式背后的原理以及封装思路，就需要对 Spring Security 的 OAuth Client 源码有一定的了解。

当 Spring Boot 2.0 工程引入 Spring Security 关于 OAuth Client 的依赖包时，Spring Boot 的自动配置策略会在 Spring Security 的过滤链中插入专门用于处理 OAuth Client 逻辑的过滤器，具体可以参考 org.springframework.boot.autoconfigure.security 包中的自动配置类。

当@EnableWebSecurity(debug = true)的 debug 属性被设置为 true 时，控制台日志将打印每一条 URL 所经过的过滤器链。

```
Security filter chain: [
 WebAsyncManagerIntegrationFilter
 SecurityContextPersistenceFilter
 HeaderWriterFilter
 CsrfFilter
 LogoutFilter
 OAuth2AuthorizationRequestRedirectFilter
 OAuth2LoginAuthenticationFilter
 DefaultLoginPageGeneratingFilter
 RequestCacheAwareFilter
 SecurityContextHolderAwareRequestFilter
 AnonymousAuthenticationFilter
 SessionManagementFilter
 ExceptionTranslationFilter
 FilterSecurityInterceptor
]
```

OAuth 客户端认证流程涉及以下三个特有的核心过滤器。

◎　OAuth2AuthorizationRequestRedirectFilter：通过重定向到 authorization-uri 端口地址来

启动授权码（或隐式授权码）模式流程，获取 code。
- OAuth2LoginAuthenticationFilter：核心 OAuth 登录过滤器，首先从 URL 中提取 code，然后使用 code 获取 access_token，接着借助 access_token 获取用户信息，最终构建出 OAuth2AuthenticationToken 认证对象，表明认证成功。
- DefaultLoginPageGeneratingFilter：用于生成默认登录页。

这三个过滤器是实现 OAuth 客户端流程的核心逻辑入口。

14.3.1　OAuth2AuthorizationRequestRedirectFilter

核心处理逻辑在 doFilterInternal 方法中。

```
@Override
protected void doFilterInternal(HttpServletRequest request,
HttpServletResponse response, FilterChain filterChain)
            throws ServletException, IOException {
  // 如果是发起授权码模式的请求
    if (this.shouldRequestAuthorization(request, response)) {
            try {
                //那么重定向请求OAuth服务提供商提供的获取code的接口
                this.sendRedirectForAuthorization(request,
response);
            } catch (Exception failed) {
                this.unsuccessfulRedirectForAuthorization(request,
response, failed);
            }
            return;
        }

        filterChain.doFilter(request, response);
}
```

其中，sendRedirectForAuthorization 方法是发送请求的具体逻辑。

```
private void sendRedirectForAuthorization(HttpServletRequest request,
HttpServletResponse response)
            throws IOException, ServletException {

    ...
    if
(AuthorizationGrantType.AUTHORIZATION_CODE.equals(authorizationRequest.getGr
antType())) {
```

```
            this.authorizationRequestRepository.saveAuthorizationRequest(author
izationRequest, request, response);
        }
        ...
}
```

OAuth2AuthorizationRequest 对象表示 OAuth 的授权请求，OAuth2AuthorizationRequest
RedirectFilter 把授权请求缓存在 authorizationRequestRepository 中，并重定向到 authorization-uri，
之后的逻辑便转交给过滤器 OAuth2LoginAuthenticationFilter。

14.3.2　OAuth2LoginAuthenticationFilter

从接收 code 到构建 OAuth2AuthenticationToken 对象，OAuth2LoginAuthenticationFilter 过滤器承载了最核心的 OAuth 客户端认证逻辑，其中，attemptAuthentication 方法如下。

```
@Override
public Authentication attemptAuthentication(HttpServletRequest request,
HttpServletResponse response)
            throws AuthenticationException, IOException,
ServletException {
        // 如果 URL 路径中没有 code 和 state 参数，或者有 error 参数，则说明获取 code 失败，
        // 直接抛出异常
        if (!this.authorizationResponseSuccess(request)
&& !this.authorizationResponseError(request)) {
            OAuth2Error oauth2Error = new
OAuth2Error(OAuth2ErrorCodes.INVALID_REQUEST);
            throw new OAuth2AuthenticationException(oauth2Error,
oauth2Error.toString());
        }
        // 从 authorizationRequestRepository 中取出在
        // OAuth2AuthorizationRequestRedirectFilter 中存入的
        // OAuth2AuthorizationRequest 对象
        OAuth2AuthorizationRequest authorizationRequest =
this.authorizationRequestRepository.loadAuthorizationRequest(request);
        if (authorizationRequest == null) {
            OAuth2Error oauth2Error = new
OAuth2Error(AUTHORIZATION_REQUEST_NOT_FOUND_ERROR_CODE);
            throw new OAuth2AuthenticationException(oauth2Error,
oauth2Error.toString());
        }
        this.authorizationRequestRepository.removeAuthorizationRequest(requ
```

```java
est);

        String registrationId = (String) 
authorizationRequest.getAdditionalParameters().get(OAuth2ParameterNames.REGI
STRATION_ID);
        ClientRegistration clientRegistration = 
this.clientRegistrationRepository.findByRegistrationId(registrationId);
        if (clientRegistration == null) {
            OAuth2Error oauth2Error = new 
OAuth2Error(CLIENT_REGISTRATION_NOT_FOUND_ERROR_CODE,
                    "Client Registration not found with Id: " + 
registrationId, null);
            throw new OAuth2AuthenticationException(oauth2Error, 
oauth2Error.toString());
        }
        // 正式获取 code
        OAuth2AuthorizationResponse authorizationResponse = 
this.convert(request);

        // 构建请求 access_token 的请求对象
        OAuth2LoginAuthenticationToken authenticationRequest = new 
OAuth2LoginAuthenticationToken(
                clientRegistration, new 
OAuth2AuthorizationExchange(authorizationRequest, authorizationResponse));
        authenticationRequest.setDetails(this.authenticationDetailsSource.b
uildDetails(request));

        //通过 code 获取 access_token, 通过 access_token 获取用户信息,
        //构建 OAuth2LoginAuthenticationToken 认证对象
        OAuth2LoginAuthenticationToken authenticationResult =
                (OAuth2LoginAuthenticationToken) 
this.getAuthenticationManager().authenticate(authenticationRequest);

        // 将 OAuth2LoginAuthenticationToken 转化为
        //OAuth2AuthenticationToken 认证对象, 融入 Spring Security 上下文逻辑
        OAuth2AuthenticationToken oauth2Authentication = new 
OAuth2AuthenticationToken(
                authenticationResult.getPrincipal(),
                authenticationResult.getAuthorities(),

        authenticationResult.getClientRegistration().getRegistrationId());

        OAuth2AuthorizedClient authorizedClient = new OAuth2AuthorizedClient(
```

```
                authenticationResult.getClientRegistration(),
                oauth2Authentication.getName(),
                authenticationResult.getAccessToken());

        this.authorizedClientService.saveAuthorizedClient(authorizedClient,
oauth2Authentication);

        return oauth2Authentication;
}
```

需要注意的是，使用 code 交换 access_token 以及通过 access_token 获取用户信息的逻辑并不在此处实现。

```
OAuth2LoginAuthenticationToken authenticationResult =
(OAuth2LoginAuthenticationToken)
this.getAuthenticationManager().authenticate(authenticationRequest);
```

getAuthenticationManager()获取的是 Spring Security 认证管理器，它通过适配器模式调用多个 Provider 对象进行认证，其中包含 OAuth 认证逻辑所对应的 Provider：OAuth2LoginAuthenticationProvider。

14.3.3　DefaultLoginPageGeneratingFilter

当没有配置自定义登录页时，自动配置机制会将 DefaultLoginPageGeneratingFilter 插到 Spring Security 过滤器链中，在合适的时机生成一个默认登录视图。

```
public void doFilter(ServletRequest req, ServletResponse res, FilterChain chain)
            throws IOException, ServletException {
        HttpServletRequest request = (HttpServletRequest) req;
        HttpServletResponse response = (HttpServletResponse) res;

        boolean loginError = isErrorPage(request);
        boolean logoutSuccess = isLogoutSuccess(request);
        if (isLoginUrlRequest(request) || loginError || logoutSuccess)
{
            String loginPageHtml = generateLoginPageHtml(request,
loginError,
                    logoutSuccess);
            response.setContentType("text/html;charset=UTF-8");

    response.setContentLength(loginPageHtml.getBytes(StandardCharsets.UTF_8).length);
```

```
                response.getWriter().write(loginPageHtml);

                return;
        }

        chain.doFilter(request, response);
}
```

14.3.4　OAuth2LoginAuthenticationProvider

OAuth2LoginAuthenticationProvider 主要实现两个核心逻辑：通过 code 交换 access_token，以及通过 access_token 获取用户信息。主要代码在 authenticate 方法中。

```
@Override
public Authentication authenticate(Authentication authentication) throws
AuthenticationException {
        OAuth2LoginAuthenticationToken authorizationCodeAuthentication =
                (OAuth2LoginAuthenticationToken) authentication;

        ...

        // 验证获取的授权码是否合法
        OAuth2AuthorizationRequest authorizationRequest =
authorizationCodeAuthentication
                .getAuthorizationExchange().getAuthorizationRequest();
        OAuth2AuthorizationResponse authorizationResponse =
authorizationCodeAuthentication
                .getAuthorizationExchange().getAuthorizationResponse();

        if (authorizationResponse.statusError()) {
                throw new OAuth2AuthenticationException(
                        authorizationResponse.getError(),
authorizationResponse.getError().toString());
        }

        if
(!authorizationResponse.getState().equals(authorizationRequest.getState()))
{
                OAuth2Error oauth2Error = new
OAuth2Error(INVALID_STATE_PARAMETER_ERROR_CODE);
                throw new OAuth2AuthenticationException(oauth2Error,
oauth2Error.toString());
        }
```

```
        if 
(!authorizationResponse.getRedirectUri().equals(authorizationRequest.getRedi
rectUri())) {
            OAuth2Error oauth2Error = new 
OAuth2Error(INVALID_REDIRECT_URI_PARAMETER_ERROR_CODE);
            throw new OAuth2AuthenticationException(oauth2Error, 
oauth2Error.toString());
        }

        // 如果授权码合法,则请求 access_token
        OAuth2AccessTokenResponse accessTokenResponse =
                this.accessTokenResponseClient.getTokenResponse(
                    new OAuth2AuthorizationCodeGrantRequest(
                        
authorizationCodeAuthentication.getClientRegistration(),
                        
authorizationCodeAuthentication.getAuthorizationExchange()));

        OAuth2AccessToken accessToken = accessTokenResponse.getAccessToken();

        // 通过 access_token 请求用户信息
        OAuth2User oauth2User = this.userService.loadUser(
                new 
OAuth2UserRequest(authorizationCodeAuthentication.getClientRegistration(),
accessToken));
        ...
}
```

14.4 Spring Security OAuth 授权服务器

很多时候,交换 QQ、微信或新浪微博等社交账号的意义甚至超越了交换名片和个人电话,因此越来越多的企业应用选择通过接入这些社交平台的 OAuth 服务来降低"获客"成本,同时为用户提供更为便捷的登录体验。

企业在发展壮大过程中,搭建自有开放平台的优势如下。

◎ 为 App 端提供统一的接口管控平台。
◎ 为第三方合作伙伴的业务对接提供授信可控的技术对接平台。
◎ 搭建基于 API 的生态体系。

OAuth 作为开放平台的代表，在企业级应用开发中占有举足轻重的地位，每个开发者都应当对它有所了解，不仅包括前面介绍的 OAuth 客户端实现，还包括授权服务器。

14.4.1 功能概述

授权服务器主要提供 OAuth Client 注册、用户认证、token 分发、token 验证、token 刷新等功能。资源服务器则负责处理具体的鉴权逻辑，包括基于 scope 的鉴权、基于配置的 Client 的 authorities 属性的鉴权，以及基于用户角色的鉴权等。

授权服务器和资源服务器既可以在一个应用内同时实现，也可以拆分成两个独立的应用。企业中常见的基于授权码模式的架构，一般是由一个授权服务器（本身也是资源服务器）和多个资源服务器组成的，如图 14-7 所示。

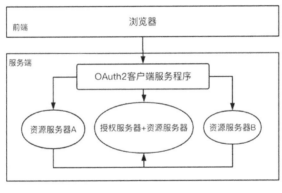

图 14-7

14.4.2 依赖包说明

Spring Security 有一个独立的子项目 spring-security-oauth，提供授权服务器和资源服务器功能。

Spring Security 内嵌的 OAuth 模块从 Spring Security 5.3 版本开始提供授权服务器功能。

而另一个项目 spring-security-oauth2-boot 则专门为在 Spring Boot 2.0 上使用 spring-security-oauth 的场景做适配。在 spring-security-oauth2-boot 的帮助下，基于 Spring Boot 2.0 搭建 OAuth 授权服务器将变得非常简便，基本依赖如下。

```
<!--oauth2-->
<dependency>
```

```xml
    <groupId>org.springframework.boot</groupId>
    <artifactId>spring-boot-starter-security</artifactId>
</dependency>
<dependency>
    <groupId>org.springframework.security.oauth.boot</groupId>
    <artifactId>spring-security-oauth2-autoconfigure</artifactId>
    <version>2.0.8.RELEASE</version>
</dependency>
```

14.4.3　编码实现

1. 新建工程

新建一个 Spring Boot 2.0 工程，命名为 auth-server，pom 文件的主要依赖包如下。

```xml
<dependency>
    <groupId>org.springframework.boot</groupId>
    <artifactId>spring-boot-starter-web</artifactId>
</dependency>
<!--oauth2-->
<dependency>
    <groupId>org.springframework.boot</groupId>
    <artifactId>spring-boot-starter-security</artifactId>
</dependency>
<dependency>
    <groupId>org.springframework.security.oauth.boot</groupId>
    <artifactId>spring-security-oauth2-autoconfigure</artifactId>
    <version>2.0.8.RELEASE</version>
</dependency>
<dependency>
    <groupId>org.springframework.boot</groupId>
    <artifactId>spring-boot-starter-test</artifactId>
    <scope>test</scope>
</dependency>
<dependency>
    <groupId>org.springframework.security</groupId>
    <artifactId>spring-security-test</artifactId>
    <scope>test</scope>
</dependency>
```

2. 开启授权服务器功能

与 Spring Boot 中一系列以 @Enable 为前缀的注解一样，只需在特定的配置类上加

@EnableAuthorizationServer 注解就可以了。

```
@Profile("minimal")
@SpringBootApplication
@EnableAuthorizationServer
public class AuthServerApplicationMinimal {
   public static void main(String[] args) {
       SpringApplication.run(AuthorizationserverApplicationMinimal.class, args);
   }
}
```

为了全方位展示授权服务器的功能，我们需要在一个工程中提供多套配置。Spring Boot 的 Profile 机制可以帮助我们快速切换到不同的环境配置。

与@EnableWebSecurity 类似，@EnableAuthorizationServer 注解会自动导入 OAuth 相关的配置类，并由这些配置类提供绝大多数的默认配置，比如 token 的签名方式、有效时间和授权类型等。

```
@Target({ElementType.TYPE})
@Retention(RetentionPolicy.RUNTIME)
@Documented
@Import({AuthorizationServerEndpointsConfiguration.class,
AuthorizationServerSecurityConfiguration.class})
public @interface EnableAuthorizationServer {
}
```

3. 注册 OAuth 客户端（Client）

在授权服务器中，至少需要注册一个客户端与授权服务器进行交互。客户端可以直接在配置文件中注册。

（1）application-minimal.yml。

```
security:
  oauth2:
    client:
      client-id: client
      client-secret: client
```

（2）application.yml。

```
server:
  port: 9999
spring:
  profiles:
    active: minimal
```

把被激活的 profile 设定为 minimal。

4. 效果演示

客户端默认只支持客户端模式（client credentials），可以在命令行窗口执行以下命令获取 access_token。

```
curl client:client@localhost:9999/oauth/token
-dgrant_type=client_credentials -dscope=any
```

返回结果如下。

```
{
    "access_token" : "f05a1ea7-4c80-4583-a123-dc7a99415588",
    "token_type" : "bearer",
    "expires_in" : 43173,
    "scope" : "any"
}
```

携带 access_token 便可以访问任何支持 OAuth 不透明令牌（Opaque Token），且通过此授权服务器进行授权管理的资源服务器。

14.5　OAuth 授权服务器功能扩展和自定义配置

前面基于 Spring Boot 提供的自动配置机制，快速搭建了一个简单的 OAuth 授权服务器。Spring Boot 默认配置一个 AuthorizationServerConfigurer 类的实例，该实例主要提供以下配置。

（1）默认的密码编码器 NoOpPasswordEncoder。在实际应用中，需要使用 BCryptPasswordEncoder 替换 NoOpPasswordEncoder 来提高密码安全性。

（2）OAuth 客户端默认支持多种授权模式。具体的授权模式有授权码模式（authorization_code）、密码模式（password）、客户端模式（client_credentials）、隐式授权模式（implicit）和刷新 token 模式（refresh_token，Spring Security 将刷新 access_token 的过程当作一种授权模式处理）。

（3）其他关键类的实例如下。

- AuthenticationManager：该实例用来提供查询用户信息的能力（注意不是查询 OAuth 客户端）。
- TokenStore：该实例用来提供生成、查询 token 的能力。
- AccessTokenConverter：该实例用来转换 access_token 的格式，默认是 Opaque 格式，可以替换为 JWT 或其他自定义格式。

然而在实际应用中，默认配置通常很难胜任各类复杂的需求场景。例如，注册多个 OAuth 客户端、不同 OAuth 客户端使用不同授权模式、更换密码编码方式等，都需要对 AuthorizationServerConfigurer 进行自定义配置。Spring Security OAuth 提供了简单的自定义配置方式，只需新建一个继承 AuthorizationServerConfigurerAdapter 的子类，再进行简单配置即可。

14.5.1 自定义配置的授权服务器

下面扩展 authorizationserver 工程，并对其进行自定义配置。为了避免与最简 OAuth 授权服务器中的类产生冲突，这里定义的所有类都将增加@Profile("simple")注解。

（1）新建程序入口类。

```
@Profile("simple")
@SpringBootApplication
public class AuthorizationserverApplicationSimple {

    public static void main(String[] args) {
        SpringApplication.run(AuthorizationserverApplicationSimple.class, args);
    }

}
```

该入口类去掉了@EnableAuthorizationServer 注解。

（2）新建授权服务器配置类。

新建一个继承自 AuthorizationServerConfigurerAdapter 的配置类 AuthorizationServerConfig，它是自定义配置的最核心入口类。

```
@Profile("simple")
@Configuration
```

```java
@EnableAuthorizationServer
public class AuthorizationServerConfig extends
AuthorizationServerConfigurerAdapter {

    @Autowired
    private PasswordEncoder passwordEncoder;

    /**
     * 配置授权服务器的安全，意味着/oauth/token 端点和/oauth/authorize 端点都应该是安全的。
     * 默认的设置覆盖了绝大多数需求，所以一般情况下不需要做任何事情
     */
    @Override
    public void configure(AuthorizationServerSecurityConfigurer security) throws Exception {
        super.configure(security);
    }

    /**
     * 该处通过配置 ClientDetailsService 来配置注册到该授权服务器的客户端 clients 信息。
     * 注意，除非在下面的 configure(AuthorizationServerEndpointsConfigurer) 中指定
     * 了一个 AuthenticationManager，否则密码授权模式不可用
     *
     * 至少要配置一个 client，否则无法启动服务器
     */
    @Override
    public void configure(ClientDetailsServiceConfigurer clients) throws Exception {
        clients.inMemory()
                // client_id
                .withClient("client-for-server")
                // client_secret
                .secret(passwordEncoder.encode("client-for-server"))
                // 该 client 支持的授权模式。OAuth 的 client 在请求 code 时，
                //只有传递授权模式参数，该处包含的授权模式才可以访问
                .authorizedGrantTypes("authorization_code", "implicit")
                //该 client 分配的 access_token 的有效时间要少于刷新时间
                .accessTokenValiditySeconds(7200)
                //该 client 分配的 access_token 的可刷新时间要多于有效时间
                //超过有效时间，但在可刷新时间范围内的 access_token 也可以刷新
                .refreshTokenValiditySeconds(72000)
                // 重定向 URL
                .redirectUris("http://localhost:8080/login/oauth2/code/authorizationserver")
```

```
            .additionalInformation()
            // 该 client 可以访问的资源服务器 ID, 每个资源服务器都有一个 ID
            .resourceIds(ResourceServerConfig.RESOURCE_ID)
            // 该 client 拥有的权限, 资源服务器可以依据该处定义的权限
            //对 client 进行鉴权
            .authorities("ROLE_CLIENT")
            // 该 client 可以访问的资源的范围, 资源服务器可以依据该处定义
            // 的范围对 client 进行鉴权
            .scopes("profile", "email", "phone", "aaa")
            // 自动批准的范围(scope), 自动批准的 scope 在批准页不需要
            //显示, 即不需要用户确认批准。如果所有 scope 都自动批准, 则不显示批准页
            .autoApprove("profile");
    }

    /**
     * 该方法是用来配置授权服务器特性的(Authorization Server endpoints), 主要是一些
     * 非安全的特性, 比如 token 存储、token 自定义、授权模式等。
     * 默认不需要任何配置, 如果需要密码授权, 则需要提供一个 AuthenticationManager
     */
    @Override
    public void configure(AuthorizationServerEndpointsConfigurer endpoints)
throws Exception {
        super.configure(endpoints);
    }

}
```

前面去掉的@EnableAuthorizationServer 注解需要添加在 AuthorizationServerConfig 上。AuthorizationServerConfig 可以提供以下三个方面的配置。

（1）ClientDetailsServiceConfigurer：配置注册的 OAuth 客户端信息，支持指定到 JDBC 数据库。

（2）AuthorizationServerSecurityConfigurer：定义 token 端点的安全策略。

（3）AuthorizationServerEndpointsConfigurer：定义获取 code 的端点（Authorization Endpoint）和获取 access_token 的端点（Token Endpoints），定义 token services。

需要注意的是，为了测试方便，目前只在内存中注册 OAuth 客户端信息，但在生产环境中，通常会使用数据库来存储 OAuth 客户端信息，这部分内容将在后文介绍。

OAuth 客户端有很多属性，这些属性代表了授权服务器的不同功能，充分理解这些属性有

助于更全面地认识授权服务器。

（1）client_id：客户端 ID，OAuth 客户端的唯一标志，功能类似于 QQ 登录的 appId。在注册 OAuth 客户端时，由客户端填写或由服务器生成，不可以为 null。

（2）client_secret：密钥。在注册 OAuth 客户端时，由客户端填写或由服务器生成，不可以为 null。

（3）authorizedgranttypes：OAuth 客户端的授权模式，可选值包括 authorization_code、password、refresh_token、implicit 和 client_credentials。在注册 OAuth 客户端时，由服务器指定，不可以为 null。如果 OAuth 客户端同时支持多种授权模式，则多种授权模式之间用英文逗号分隔，比如，同时支持授权码类型和密码类型："authorization_code,password"。

（4）access_token_validity：OAuth 客户端的 access_token 的有效时间（单位为秒）。在注册 OAuth 客户端时由客户端填写或由服务器生成，可以为 null。如果该值为 null，则系统会使用默认值（60×60×12，即 12 小时）。

（5）refresh_token_validity：OAuth 客户端的 refresh_token 的有效时间（单位为秒）。在注册 OAuth 客户端时由客户端填写或由服务器生成，可以为 null。如果该值为 null，则系统会使用默认值（60×60×24×30，即 30 天）。设置该值时需要注意 refresh_token_validity 一定要大于 access_token_validity，才能保证 access_token 可以正常刷新。

（6）redirect_uri：重定向 URI。注册 OAuth 客户端时由客户端填写，不可以为 null。当为授权码模式或隐式授权模式时，授权服务器通过携带 code 重定向到该 URI 的方式，将 code 传递给 OAuth 客户端。在整个 OAuth 的流程中，有很多步骤可用来判断该值的正确性。

（7）授权码模式。

- 当获取 code 时：在客户端发起的请求中必须携带 redirect_uri 参数，该参数的值必须与客户端注册的 redirect_uri 值一致。
- 当用 code 换取 access_token 时：客户端必须传递相同的 redirect_uri 值。
- 隐式授权模式：通过 redirect_uri 的 hash 值来传递 access_token 值。

（8）additionalInformation：预留的字段。在注册 OAuth 客户端时由客户端填写，可以为 null。该属性值必须是 JSON 格式的字符串。在实际应用中，可以用该字段存储客户端的其他信息，如客户端的国家、地区、注册时的 IP 地址等。

（9）resource_ids：OAuth 客户端所能访问的资源服务器的 id 集合。注册 OAuth 客户端时由客户端填写，可以为 null。

（10）authorities：OAuth 客户端拥有的 Spring Security 的权限值。在注册 OAuth 客户端时由客户端填写，可以为 null。如果有多个权限值，则用英文逗号分隔，比如，"ROLE_UNITY,ROLE_USER"。需要根据不同的授权模式来判断是否设置该字段的值。

（11）密码模式或授权码模式：不需要指定该字段，因为服务器将根据用户在服务器中拥有的权限来判断用户是否有权限访问对应的 API。

（12）隐式授权模式或客户端模式：需要指定该字段，因为服务器将根据该字段的内容来判断用户是否有权限访问对应的 API。

（13）scope：OAuth 客户端申请的权限范围，比如，我们在 QQ 登录时配置的"scope:get_user_info,list_album,upload_pic,do_like"。在注册 OAuth 客户端时由客户端填写或由服务器生成，不可以为 null。

（14）autoapprove：用户是否自动配置，默认为"false"，可选值为 scope 属性中定义的值。该字段只适用于授权码模式。在注册 OAuth 客户端时由客户端填写或由服务器生成，可以为 null。当用户登录成功后，若该值设置为"true"或其他支持的 scope 值，则会跳过用户批准页，自动获得授权。

1. 新建资源服务配置类

当使用授权码模式时，OAuth 客户端在获取 access_token 后，会从后台调用授权服务器的用户信息端点来构建认证对象（Authentication）。因为用户信息端点需要被资源服务器保护起来，所以需要将授权服务器同时配置为资源服务器。

Spring Security OAuth 的资源服务器通过 Bearer 类型的 token 来保护资源端点，token 支持 JWT 和 Opaque 两种格式。只需使用 @EnableResourceServer 注解便可以将 Spring Boot 2.0 工程配置为资源服务器，并且只需新建一个 ResourceServerConfigurerAdapter 的子类，便可以开启资源服务器的自定义配置。

```
/**
 * 资源服务器的职责是对来自 OAuth 客户端的 access_token 进行鉴权。一个资源服务器包含多个端
 * 点（接口），一部分端点作为资源服务器的资源提供给 OAuth 的 client 访问，另一部分端点不由资
 * 源服务器管理。由资源服务器管理的端点安全性配置在此类中，其余端点的安全性配置在
```

```
 * SecurityConfiguration 类中。当请求中包含 OAuth2 access_token 时, Spring Security
 * 会根据资源服务器配置进行过滤。EnableResourceServer 会创建一个
 * WebSecurityConfigurerAdapter, 执行顺序 (Order) 是 3。在 SecurityConfiguration 类
 * 之前运行, 优先级更高
 */
@Profile("simple")
@Configuration
@EnableResourceServer
public class ResourceServerConfig extends ResourceServerConfigurerAdapter {

    private static final Logger logger =
LoggerFactory.getLogger(ResourceServerConfig.class);

    public static final String RESOURCE_ID = "authorizationserver";

    @Override
    public void configure(ResourceServerSecurityConfigurer resources) throws
Exception {
        super.configure(resources);
        resources.resourceId(RESOURCE_ID);
    }

    @Override
    public void configure(HttpSecurity http) throws Exception {
        logger.info("ResourceServerConfig 中配置 HttpSecurity 对象执行");
        // 只有/me 端点作为资源服务器的资源
        http.requestMatchers().antMatchers("/me")
                .and()
                .authorizeRequests()
                .anyRequest().authenticated();

    }
}
```

ResourceServerConfigurerAdapter 提供了以下两方面的自定义配置。

（1）HttpSecurity：Spring Security 的安全配置类，注意此处配置的 URL 与过滤器映射关系的优先级要高于 SecurityConfiguration 类。

（2）ResourceServerSecurityConfigurer：配置 ResourceServerTokenServices、resourceId 等。

ResourceServerTokenServices 的配置很关键。OAuth 的资源服务器主要负责对传递进来的

access_token 进行验证，验证通过后才允许 OAuth 客户端访问资源端点。在实际应用中，由于授权服务器和资源服务器的系统架构不同，所以具体的验证方式也不同，常用的验证方式有以下三种。

（1）当授权服务器与资源服务器在同一个应用中时，使用默认的 DefaultTokenServices 在服务器内部进行验证。

（2）当授权服务器和资源服务器是分离的两个应用，且 access_token 类型为 Opaque 时，使用 RemoteTokenServices 资源服务器远程调用授权服务器进行验证。

（3）当授权服务器和资源服务器是分离的两个应用，且 access_token 类型为 JWT 时，一般使用 DefaultTokenServices 资源服务器远程调用授权服务器进行验证。注意，需要将 DefaultTokenServices 的 tokenStore 属性设置为 JwtTokenStore。

另外，需要配置属性 resourceId 为"authorizationserver"，代替默认值"oauth2-resource"。

2. 配置 Spring Security

SecurityConfiguration 继承自 WebSecurityConfigurerAdapter，是 Spring Security 的标准配置入口。在此项目中，主要用来配置 PasswordEncoder、默认的用户（user/user 和 admin/admin）和 HttpSecurity 安全规则。

```
@Profile("simple")
@Configuration
@EnableWebSecurity(debug = true)
public class SecurityConfiguration extends WebSecurityConfigurerAdapter {

    private static final Logger logger =
LoggerFactory.getLogger(SecurityConfiguration.class);

    @Autowired
    public void globalUserDetails(AuthenticationManagerBuilder auth) throws Exception {
        auth.inMemoryAuthentication()
                .withUser("user").password(passwordEncoder().encode("user")).roles("USER")
                .and().withUser("admin").password(passwordEncoder().encode("admin")).roles("ADMIN");
    }
```

```java
@Override
protected void configure(HttpSecurity http) throws Exception {
    logger.info("SecurityConfiguration 中配置 HttpSecurity 对象执行");

    http.authorizeRequests()
            .antMatchers("/").permitAll()
            .anyRequest().hasAnyRole("USER", "ADMIN")
        .and().formLogin();
}

@Override
public void configure(WebSecurity web) throws Exception {
    super.configure(web);
    web.ignoring().antMatchers("/favicon.ico") ;
}

@Bean
public PasswordEncoder passwordEncoder() {
    return new BCryptPasswordEncoder();
}
}
```

至此，我们已经自定义了以下三个配置相关的类。

（1）AuthorizationServerConfig：授权服务器配置。

（2）ResourceServerConfig：资源服务器配置。

（3）SecurityConfiguration：Spring Security 安全规则配置。

注意：ResourceServerConfig 比 SecurityConfiguration 先执行，并且配置优先级更高。在控制台打印日志时，显示如图 14-8 所示。

```
.s.o.p.e.FrameworkEndpointHandlerMapping : Mapped "{[/oauth/token],methods=[POST]}" onto public org
.s.o.p.e.FrameworkEndpointHandlerMapping : Mapped "{[/oauth/check_token]}" onto public java.util.Ma
.s.o.p.e.FrameworkEndpointHandlerMapping : Mapped "{[/oauth/confirm_access]}" onto public org.sprin
.s.o.p.e.FrameworkEndpointHandlerMapping : Mapped "{[/oauth/error]}" onto public org.springframewor
o.a.conifg.ResourceServerConfig          : ResourceServerConfig中配置HttpSecurity对象执行
o.a.conifg.SecurityConfiguration         : SecurityConfiguration中配置HttpSecurity对象执行
o.s.s.web.DefaultSecurityFilterChain     : Creating filter chain: OrRequestMatcher [requestMatchers
o.s.s.web.DefaultSecurityFilterChain     : Creating filter chain: OrRequestMatcher [requestMatchers
o.s.s.web.DefaultSecurityFilterChain     : Creating filter chain: org.springframework.security.web.
```

图 14-8

为了保证不同端点对应的过滤器链不冲突，在配置 ResourceServerConfig 时，需要对不同类型的端点有清晰的规划，分辨出哪些端点是被资源服务器保护的资源端点，哪些端点是被

Spring Security 默认安全机制保护的普通端点。

3. 编写测试路由

在前面的步骤中，关于配置的部分已经完成，下面定义一些测试路由来进行验证。首先，定义一些普通接口。

```java
@Profile("simple")
@RestController
public class MainController {

    @GetMapping("/")
    public String email() {
        return "这是主页";
    }

    @GetMapping("/admin")
    public String admin() {
        return "这是 admin 页";
    }

    @GetMapping("/user")
    public String user() {
        return "这是 user 页";
    }

}
```

其次，定义一些由资源服务器保护的资源端点（实际上也是接口）。

```java
/**
 * 获得认证信息，当认证通过后，第三方应用可以请求的资源
 */
@Profile("simple")
@RestController
public class ResourceController {

    private static final Logger logger = LoggerFactory.getLogger(ResourceController.class);

    @RequestMapping("/me")
    public Principal me(Principal principal) {
        logger.debug(principal.toString());
```

```
        return principal;
    }
}
```

此处提供了用户信息端点"/me"，供 OAuth 客户端获取认证对象时调用。

4. 运行授权服务器

将项目配置文件中活动的 profile 设置为"simple"，并启动授权服务器。

```
spring:
  profiles:
    active: simple
```

根据配置，当用户在未登录的情况下访问 http://localhost:9999/admin 时会跳转到登录页。在用户输入 user/user 或 admin/admin 登录成功后，即可访问数据。如果直接访问 http://localhost:9999/me 端点，则会报错，如图 14-9 所示。

```
This XML file does not appear to have any style information associated with it. The document tree is shown below.

▼<oauth>
 ▼<error_description>
    Full authentication is required to access this resource
  </error_description>
  <error>unauthorized</error>
</oauth>
```

图 14-9

通过查看控制台日志，可以看到"/me"的过滤器链。

```
Security filter chain: [
 WebAsyncManagerIntegrationFilter
 SecurityContextPersistenceFilter
 HeaderWriterFilter
 LogoutFilter
 OAuth2AuthenticationProcessingFilter
 RequestCacheAwareFilter
 SecurityContextHolderAwareRequestFilter
 AnonymousAuthenticationFilter
 SessionManagementFilter
 ExceptionTranslationFilter
 FilterSecurityInterceptor
]
```

与其他端点的过滤器链稍有不同。

```
Security filter chain: [
 WebAsyncManagerIntegrationFilter
 SecurityContextPersistenceFilter
 HeaderWriterFilter
 CsrfFilter
 LogoutFilter
 RequestCacheAwareFilter
 SecurityContextHolderAwareRequestFilter
 AnonymousAuthenticationFilter
 SessionManagementFilter
 ExceptionTranslationFilter
 FilterSecurityInterceptor
]
```

因为"/me"端点只作为 OAuth 的资源端点为 OAuth 客户端提供服务,所以它比普通接口多了一个 OAuth2AuthenticationProcessingFilter 过滤器,这个过滤器正是资源服务器工作流程的核心入口。

至此就完成了 OAuth 授权服务器的功能扩展和自定义配置。下面将专门构建一个 OAuth 客户端程序来访问授权服务器。

14.5.2 编写 OAuth 客户端

1. 初始化工程

参考前面介绍的 OAuth 客户端工程 client。复制 client 工程,将其重命名为 client-for-server,并对项目配置文件做一些修改。

```
spring:
  security:
    oauth2:
      client:
        registration:
          authorizationserver:
            client-id: client-for-server
            client-secret: client-for-server
            provider: authorizationserver
            client-name: 使用自定义授权服务器登录
            authorization-grant-type: authorization_code
            client-authentication-method: basic
```

```
          scope: profile,email,phone
          redirect-uri-template:
"{baseUrl}/login/oauth2/code/{registrationId}"
    provider:
      authorizationserver:
        authorization-uri: http://localhost:9999/oauth/authorize
        token-uri: http://localhost:9999/oauth/token
        user-info-uri: http://localhost:9999/me
        user-name-attribute: "name"
```

其中，registration 的 id 需要设置为 authorizationserver，即我们构建的授权服务器。client-id 为 client-for-server，其余配置信息与授权服务器中注册的客户端配置信息保持一致即可。

2. 效果演示

- 启动授权服务器 authorizationserver。
- 启动 OAuth 客户端 client-for-server。
- 当用浏览器访问 client-for-server 客户端地址 http://localhost:8080 时，浏览器将自动跳转到 authorizationserver 登录页。在用户成功完成授权码流程，登录成功后，浏览器会显示字符串 "Hello..."。

14.5.3 使用 JDBC 存储 OAuth 客户端信息

OAuth 客户端信息默认存储在内存中，但在实际应用时通常需要将其存储到关系数据库中。为了便于演示，此处将使用内嵌的 H2 数据库代替一般的关系数据库。为了防止冲突，还应该将新建的相关类都使用@Profile("jdbc")进行注解。

1. 引入相关依赖

如果只是验证 JDBC，那么并不需要搭建真正的关系型数据库。实际上，内嵌的 H2 数据库即可胜任。在使用 H2 数据库时需要引入两个依赖。

```
<!--jdbc-->
<dependency>
    <groupId>org.springframework.boot</groupId>
    <artifactId>spring-boot-starter-jdbc</artifactId>
</dependency>
<dependency>
    <groupId>com.h2database</groupId>
    <artifactId>h2</artifactId>
```

```
    <scope>runtime</scope>
</dependency>
```

2. H2 数据库初始化配置

(1) 定义数据库表与初始化数据。

在使用默认配置时，Spring Boot 工程 resources 目录下的 data.sql 与 schema.sql 会自动被执行，如图 14-10 所示。

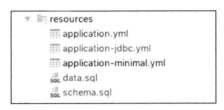

图 14-10

schema.sql 中存放的是初始化数据库表结构的 SQL 语句，在官方源码中可以看到详细的定义。

```sql
-- OAuth 客户端信息
-- 主要操作 oauth_client_details 表的类是 JdbcClientDetailsService
create table oauth_client_details (
  client_id VARCHAR(256) PRIMARY KEY,
  resource_ids VARCHAR(256),
  client_secret VARCHAR(256),
  scope VARCHAR(256),
  authorized_grant_types VARCHAR(256),
  web_server_redirect_uri VARCHAR(256),
  authorities VARCHAR(256),
  access_token_validity INTEGER,
  refresh_token_validity INTEGER,
  additional_information VARCHAR(4096),
  autoapprove VARCHAR(256)
);

-- 存储从服务器获取的 token 数据
-- 主要操作 oauth_client_token 表的类是 JdbcClientTokenServices
create table oauth_client_token (
  token_id VARCHAR(256),
  token LONGVARBINARY,
  authentication_id VARCHAR(256) PRIMARY KEY,
```

```sql
  user_name VARCHAR(256),
  client_id VARCHAR(256)
);

-- 存储 access_token
-- 主要操作 oauth_access_token 表的类是 JdbcTokenStore
create table oauth_access_token (
  token_id VARCHAR(256),
  token LONGVARBINARY,
  authentication_id VARCHAR(256) PRIMARY KEY,
  user_name VARCHAR(256),
  client_id VARCHAR(256),
  authentication LONGVARBINARY,
  refresh_token VARCHAR(256)
);

-- 存储 refresh_token
-- 主要操作 oauth_refresh_token 表的类是 JdbcTokenStore
create table oauth_refresh_token (
  token_id VARCHAR(256),
  token LONGVARBINARY,
  authentication LONGVARBINARY
);

-- 主要操作 oauth_code 表的类是 JdbcAuthorizationCodeServices
create table oauth_code (
  code VARCHAR(256), authentication LONGVARBINARY
);

-- 授权记录表
create table oauth_approvals (
        userId VARCHAR(256),
        clientId VARCHAR(256),
        scope VARCHAR(256),
        status VARCHAR(10),
        expiresAt TIMESTAMP,
        lastModifiedAt TIMESTAMP
);

-- 用于定制 oauth_client_details 表
create table ClientDetails (
  appId VARCHAR(256) PRIMARY KEY,
  resourceIds VARCHAR(256),
```

```
  appSecret VARCHAR(256),
  scope VARCHAR(256),
  grantTypes VARCHAR(256),
  redirectUrl VARCHAR(256),
  authorities VARCHAR(256),
  access_token_validity INTEGER,
  refresh_token_validity INTEGER,
  additionalInformation VARCHAR(4096),
  autoApproveScopes VARCHAR(256)
);
```

具体请参阅官方源码：https://github.com/spring-projects/spring-security-oauth/blob/master/spring-security-oauth2/src/test/resources/schema.sql。

data.sql 中存放的是初始化数据的 SQL 语句。

```
-- 注册 client-for-server 客户端
-- client_secret 使用 BCryptPasswordEncoder 加密后的字符串
insert into oauth_client_details
( 'client_id', 'resource_ids', 'client_secret', 'scope',
'authorized_grant_types',
'web_server_redirect_uri', 'authorities', 'access_token_validity',
'refresh_token_validity',
 'additional_information','autoapprove')
values
( 'client-for-server', 'authorizationserver',
'$2a$10$UmAZtVnLzegIdP7BfBEHz.f8bMqRd3ZOlFeUKBanAOGCO8iIFQaE.',
'profile,email,phone', 'authorization_code',
'http://localhost:8080/login/oauth2/code/authorizationserver',
 'ROLE_CLIENT', '7200', '72000', '{}', 'profile');
```

（2）访问 H2 数据库控制台。

H2 数据库默认提供一个可以在浏览器中访问的控制台，访问路径为 "/h2-console"。

新建 application-jdbc.yml。

```
spring:
  h2:
    console:
      # 使用 H2 数据库 控制台
      enabled: true
```

修改 SecurityConfiguration 安全规则。

```
@Profile("jdbc")
@Configuration
@EnableWebSecurity(debug = true)
public class SecurityConfiguration extends WebSecurityConfigurerAdapter {

    ...

    @Override
    protected void configure(HttpSecurity http) throws Exception {
        ...
        http.authorizeRequests()
                // 在实际项目中，这样配置不安全。为便于演示，将 H2 数据库控制台
                //设置为可以访问
                .antMatchers("/h2-console/**").permitAll();
                ...

        // 在实际项目中，这样配置不安全。为便于演示，临时禁用 H2 数据库控制台的
        // CSRF 防护
        http.csrf().ignoringAntMatchers("/h2-console/**");
        // 在实际项目中，这样配置不安全。为便于演示，临时允许同一来源
        // 的 H2 数据库控制台的请求
        http.headers().frameOptions().sameOrigin();
    }

    ...
}
```

至此，就可以直接在浏览器中输入 http://localhost:9999/h2-console 来访问 H2 数据库控制台了。

（3）配置数据源为 JDBC。

```
@Profile("jdbc")
@Configuration
@EnableAuthorizationServer
public class AuthorizationServerConfig extends
AuthorizationServerConfigurerAdapter {

    // 使用嵌入式 H2 数据库定义一个 DataSource 对象
    @Bean
    public DataSource dataSource() {
        return new EmbeddedDatabaseBuilder()
```

```
            .setType(EmbeddedDatabaseType.H2)
            .build();
    }

    ...

    // 指定 clients 的数据源为 JDBC
    @Override
    public void configure(ClientDetailsServiceConfigurer clients) throws Exception {
        clients.jdbc(dataSource());
    }

    ...
}
```

（4）效果演示。

修改授权服务器 authorizationserver 的 application.yml，启动授权服务器。

```
spring:
  profiles:
    active: jdbc
```

启动 OAuth 客户端中的 client-for-server 工程。

在浏览器中输入 http://localhost:9999/h2-console，效果如图 14-11 所示。

图 14-11

- ◎ 单击"连接"按钮，进入控制台，如图 14-12 所示。
- ◎ 在浏览器中输入 http://localhost:8080，访问 OAuth 客户端，成功执行认证流程后，浏览器上将显示字符串"Hello ..."。

图 14-12

14.5.4 使用 JDBC 存储 token

除 OAuth 客户端的存储位置可以自定义外，OAuth 的 token 存储位置也可以自定义。

默认情况下，access_token 和 refresh_token 是在内存中维护的。管理 token 存储位置的抽象接口是 TokenStore，具体实现类有 InMemoryTokenStore、JdbcTokenStore、JwtTokenStore 和 RedisTokenStore，分别对应了在内存、关系数据库、JWT Token 自身和 Redis 数据库中存储。下面使用 JDBC 来存储 token。

1. 数据库与初始化数据

由于在前面的 schema.sql 中已经定义了 token 相关表：oauth_access_token、oauth_refresh_token 等，所以这里不需要再定义了。

2. 配置授权服务器

在 AuthorizationServerConfig 中配置 token 的存储介质为 JDBC。

```
@Override
public void configure(AuthorizationServerEndpointsConfigurer endpoints) throws Exception {
    super.configure(endpoints);
```

```
    endpoints.tokenStore(jdbcTokenStore());
}
/**
 * jdbc token 配置
 */
@Bean
public TokenStore jdbcTokenStore() {
    return new JdbcTokenStore(dataSource());
}
```

3. 效果演示

- 用浏览器访问 http://localhost:8080，成功。
- 用浏览器访问控制台 http://localhost:9999/h2-console ，执行 SQL 语句。

```
SELECT * FROM OAUTH_ACCESS_TOKEN
```

如果发现已经有了一条 token 记录，则说明配置是成功的。

14.5.5 其他功能配置

Spring Security OAuth 的授权服务器扩展性非常强，几乎流程中的每个步骤都可以很方便地进行自定义。例如：

- 自定义批准页面。
- 配置 OAuth 客户端的各种属性。

对于这部分配置，建议读者自行通过修改各个属性的方式来观察各个属性的作用。

14.6 实现 OAuth 资源服务器

资源服务器（Resource Server）也被称作资源提供者，用来存放受保护的资源，并且实现对受保护资源的访问控制。其功能类似于传统的 API Server。在大规模应用部署场景下，通常需要配备多个资源服务器。Google 便拥有几十个资源服务器，如 Google 云平台、Google 地图、YouTube 等。每个资源服务器都是独立的，但它们可以共享同一个授权服务器，客户端在一个授权服务器获取 access_token 后，便可以使用这个 access_token 访问多个共享此授权服务器的资源服务器。

资源服务器使用 OAuth 的 access_token 来保护被定义为资源的端口。在授权码模式中，因为 access_token 缓存在了 OAuth 客户端后台应用而不是浏览器中，所以资源端口只能由客户端后台应用来访问，而无法由浏览器直接访问。OAuth 客户端后台应用可以通过 WebClient 或者 RestTemplate 携带 access_token 发送 HTTP 请求，访问资源服务器的资源端口。资源服务器中的任意端口都可以是资源端口。

在 Spring Security 5.1.x 版本的 OAuth 模块中增加了资源服务器子模块 oauth2-resource-server，但是其实现的资源服务器功能还不够强大，并且其本身还处于快速迭代的不稳定阶段，因此我们仍然选择稳定而强大的 spring-security-oauth 项目提供的 API。

14.6.1 依托于授权服务器的资源服务器

1. 新增资源端口

修改 authorizationserver 工程，在 ResourceController 中新增一个端口。

```java
/**
 * 获得认证信息，当认证通过后，第三方应用可以请求的资源
 */
@Profile("otherResourceServer")
@RestController
public class ResourceController {

    private static final Logger logger =
LoggerFactory.getLogger(ResourceController.class);

    @RequestMapping("/me")
    public Principal me(Principal principal) {
        logger.debug(principal.toString());
        return principal;
    }

    @GetMapping("/phone")
    public String phone() {
        return "phone: 1234567890";
    }
}
```

2. 配置资源端口

在资源服务器配置 ResourceServerConfig 中,将("/me")修改为("/me", "/phone")。

```
@Override
public void configure(HttpSecurity http) throws Exception {
    ...
    http.requestMatchers().antMatchers("/me", "/phone")
    ...
}
```

3. 访问资源端口

修改客户端 client-for-server 工程,新增 ResourceEndpointController。

```
/**
 * 使用 RestTemplate 携带的 access_token 请求资源服务器,获取信息
 */
@RestController
public class ResourceEndpointController {

    private static final String URL_GET_USER_PHONE =
"http://localhost:9999/phone";

    @Autowired
    private OAuth2AuthorizedClientService authorizedClientService;

    private RestTemplate restTemplate;

    private RestTemplate getRestTemplate() {
        if (restTemplate == null) {
            restTemplate = new RestTemplate();
        }
        return restTemplate;
    }

    @GetMapping("/phone")
    public String userinfo(OAuth2AuthenticationToken authentication) {
        OAuth2AuthorizedClient authorizedClient =
authorizedClientService.loadAuthorizedClient(
                authentication.getAuthorizedClientRegistrationId(),
authentication.getName());

        HttpHeaders header = new HttpHeaders();
```

```
        header.set("Authorization", "Bearer " +
authorizedClient.getAccessToken().getTokenValue());
        HttpEntity<String> requestEntity = new HttpEntity<String>(null, header);

        ResponseEntity<String> response =
getRestTemplate().exchange(URL_GET_USER_PHONE, HttpMethod.GET, requestEntity,
String.class);
        return response.getBody();
    }
}
```

使用单例模式自定义一个 RestTemplate 对象。当浏览器请求客户端的"/phone"端口时，客户端使用 RestTemplate 请求资源服务器接口"/phone"。关键配置如下。

```
 header.set("Authorization", "Bearer " +
authorizedClient.getAccessToken().getTokenValue());
```

在 HTTP 请求头的"Authorization"字段中放入 access_token，必须以"Bearer"（包含空格）开头，说明是 Bearer 类型的 token。

4. 效果演示

- 启动授权服务器 authorizationserver 工程。
- 启动客户端 client-for-server 工程。
- 用浏览器访问 http://localhost:8080/phone，在完成授权流程后，浏览器可成功显示字符串"phone: 1234567890"。

14.6.2　独立的资源服务器

资源服务器既可以与授权服务器一体，也可以抽离为一个独立的后端应用。将 Spring Boot 2.0 应用配置为资源服务器的方法非常简单，只需要引入相关的 jar 包，然后使用 @EnableResourceServer 注解在 ResourceServerConfigurerAdapter 的子类上，最后在配置文件中配置好 access_token 的相关验证方式即可。

1. 新建资源服务器

新建 Spring Boot 2.0 工程，命名为 resourceserver。pom 文件依赖包如下。

```
<dependency>
    <groupId>org.springframework.boot</groupId>
```

```xml
    <artifactId>spring-boot-starter-web</artifactId>
</dependency>
<!--oauth2-->
<dependency>
    <groupId>org.springframework.boot</groupId>
    <artifactId>spring-boot-starter-security</artifactId>
</dependency>
<dependency>
    <groupId>org.springframework.security.oauth.boot</groupId>
    <artifactId>spring-security-oauth2-autoconfigure</artifactId>
    <version>2.0.8.RELEASE</version>
</dependency>
<dependency>
    <groupId>org.springframework.boot</groupId>
    <artifactId>spring-boot-starter-test</artifactId>
    <scope>test</scope>
</dependency>
<dependency>
    <groupId>org.springframework.security</groupId>
    <artifactId>spring-security-test</artifactId>
    <scope>test</scope>
</dependency>
```

2. 配置资源服务器

增加配置类 ResourceServerConfig。

```
@Configuration
@EnableResourceServer
public class ResourceServerConfig extends ResourceServerConfigurerAdapter {

    private static final Logger logger =
LoggerFactory.getLogger(ResourceServerConfig.class);

    public static final String RESOURCE_ID = "resourceserver";

    @Override
    public void configure(ResourceServerSecurityConfigurer resources) throws
Exception {
        super.configure(resources);
        resources.resourceId(RESOURCE_ID);
    }

    @Override
```

```
public void configure(HttpSecurity http) throws Exception {
    logger.info("ResourceServerConfig 中配置HttpSecurity 对象执行");
    // 只有/user 端口作为资源服务器的资源
    http.requestMatchers().antMatchers("/resource")
            .and()
            .authorizeRequests()
            .anyRequest().authenticated();

}
}
```

将 resourceId 设置为 resourceserver。@EnableResourceServer 注解将自动向 Spring Security 过滤器链插入 OAuth2AuthenticationProcessingFilter。

3. 配置资源端口

增加 ResourceController。

```
/**
 * 获得认证信息，当认证通过后，第三方应用可以请求的资源
 */
@RestController
public class ResourceController {

    private static final Logger logger =
LoggerFactory.getLogger(ResourceController.class);

    @GetMapping("/resource")
    public String phone() {
        logger.info("in resource");
        return "resource";
    }

}
```

4. 修改资源服务器配置文件

这里使用的是 security.oauth2.resource.user-info-uri 属性，资源服务器在收到来自 OAuth 客户端的资源请求后，会携带客户端传来的 access_token，请求授权服务器的用户信息端口，构建认证对象，同时间接地完成对 access_token 的验证。

```
server:
  port: 9090
```

```yaml
spring:
  profiles:
    active: test
logging:
  level:
    root: INFO
    org.springframework.web: INFO
    org.springframework.security: DEBUG
    org.springframework.boot.autoconfigure: DEBUG
security:
  oauth2:
    resource:
      # token-info-uri: http://localhost:9999/oauth/check_token
      user-info-uri: http://localhost:9999/me
```

如果是 JWT 格式的 token，则使用 key-set-uri。

```yaml
spring:
  security:
    oauth2:
      resource:
        jwk:
          key-set-uri: {custom}
```

5. 配置授权服务器

修改授权服务器 authorizaserver 工程的 AuthorizationServerConfigMultiRes 类。

```java
@Override
public void configure(ClientDetailsServiceConfigurer clients) throws Exception
{
    clients.inMemory()
        ...
        // 此 client 可以访问的资源服务器 ID，每个资源服务器都有一个 ID
        .resourceIds(ResourceServerConfig.RESOURCE_ID, "resourceserver")
        ...
}
```

可以看到在支持的 resourceIds 中增加了资源服务器"resourceserver"。

6. 访问资源服务器资源端口

在客户端 client-for-server 中增加 AlongResourceEndpointController

```
@RestController
public class AlongResourceEndpointController {

    // 资源服务器 resourceserver 的 API
    private static final String URL_GET_RES= "http://localhost:9090/resource";

    @Autowired
    private OAuth2AuthorizedClientService authorizedClientService;

    private RestTemplate restTemplate;

    private RestTemplate getRestTemplate() {
        if (restTemplate == null) {
            restTemplate = new RestTemplate();
        }

        return restTemplate;
    }

    @GetMapping("/resource")
    public String userinfo(OAuth2AuthenticationToken authentication) {
        OAuth2AuthorizedClient authorizedClient = authorizedClientService.loadAuthorizedClient(
                authentication.getAuthorizedClientRegistrationId(),
authentication.getName());

        HttpHeaders header = new HttpHeaders();
        header.set("Authorization", "Bearer " + authorizedClient.getAccessToken().getTokenValue());
        HttpEntity<String> requestEntity = new HttpEntity<String>(null, header);

        ResponseEntity<String> response = getRestTemplate().exchange(URL_GET_RES, HttpMethod.GET, requestEntity, String.class);
        return response.getBody();
    }
}
```

当浏览器请求"/resource"端口时，client-for-server 会将其转发到资源服务器的资源端口"http://localhost:9090/resource"。

7. 效果演示

◎ 启动授权服务器 authorizationserver。

- ◎ 启动资源服务器 resourceserver。
- ◎ 启动客户端 client-for-client。
- ◎ 用浏览器访问 http://localhost:8080/resource，浏览器重定向到自动生成的 OAuth 登录页，如图 14-13 所示。
- ◎ 单击"使用自定义授权服务器登录"链接，跳转到授权服务器的登录页面，如图 14-14 所示。

成功后浏览器将显示字符串"resource"。至此，我们便成功构建了一个独立的资源服务器。

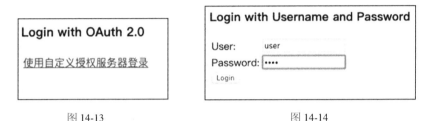

图 14-13　　　　　　　　　　　图 14-14

14.7　Spring Security OAuth 核心源码分析

虽然 Spring Security OAuth 的源码量相对较多，但得益于良好的接口设计，使得其可读性非常高。Spring Security OAuth 的源码大体可以分为两块：核心功能类和配置相关类。其中，配置相关类占绝大部分，而掌握源码"主脉"的核心功能类则相对较少。实际上，几乎所有 Spring 体系的项目结构都可以用图 14-15 来表示。

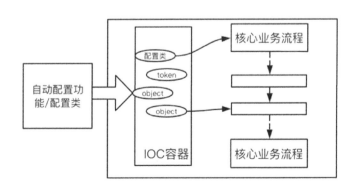

图 14-15

在类的命名上，Spring Security OAuth 遵循了非常良好的规范，例如，以 Adapter 结尾的属于适配器，以 Configurer 结尾的属于配置器，以 Builder 结尾的属于构造器，它们分别代表了不同的设计模式。正所谓"拨开云雾见青天"，忽略不影响核心逻辑的代码，寻找程序入口，跟踪运行主线程，是我们阅读开源项目的基本要领，这同样适用于 Spring Security OAuth。众所周知，Spring 体系的入口通常都是配置类，而 Spring Security OAuth 的配置类主要有三种：

- 以"Configuration"结尾的类，如 ResourceServerConfiguration、AuthorizationServerSecurityConfiguration。
- 以"Configurer"结尾的类，如 ResourceServerSecurityConfigurer、AuthorizationServerSecurityConfigurer。
- 以"ConfigurerAdapter"结尾的类，如 ResourceServerConfigurerAdapter、AuthorizationServerConfigurerAdapter。

配置类为核心业务逻辑提供实例对象，无论提供什么形式的配置入口和配置方式，其实都希望能够提供更便捷的使用方式以及更灵活的扩展能力，以真正胜任错综多变的需求场景。

14.7.1 授权服务器核心源码分析

授权服务器主要负责为客户端和资源服务器提供 access_token 服务，包括 access_token 的生成、刷新、验证、销毁等生命周期相关的功能。这些功能都是通过 HTTP 端口的方式提供的，因而源码的核心入口就是各个功能端口。

1. 源码的复杂性

如果 Spring Security OAuth 只是单纯地实现 access_token 的增、删、改、查，那么代码本身并不复杂，但作为一个完善而稳定的框架，所要考虑的问题不只是增、删、改、查这么简单。

（1）业务复杂性。

- access_token 的生成和分发：面对四种授权码模式，要充分考虑 access_token 如何持久化等复杂业务问题。
- access_token 验证：针对不同类型的 access_token，如 opaque token、JWT token 等，验证方式不同，token 的验证规则也不同。
- 签名算法：加密方式的复杂性。
- 客户端信息管理的其他问题：需要考虑各种范围权限等。

（2）与 Spring Security 体系的完美融合。

◎ 与 Spring Security 进行优雅整合的同时，完整实现 OAuth 规定的功能。

（3）扩展性。

◎ 基本属性的配置，如端点的权限、自定义的批准页面等。
◎ TokenStore 的配置，如 RedisTokenStore、JwtTokenStore、JDBCTokenStore。
◎ 其他可自定义的关键步骤。

2. 交互端口

交互端口是授权服务器一些关键流程的核心入口，授权服务器提供的交互端口有以下几个。

（1）AuthorizationEndpoint 类提供的"/oauth/authorize"端口，用于处理获取 code 的流程。

（2）TokenEndpoint 类提供的"/oauth/token"端口，用于处理获取 access_token 的流程。

（3）CheckTokenEndpoint 类提供的"/oauth/check_token"端口，用于检测 access_token 的有效性。

（4）当 access_token 为 JWT 格式时，资源服务器需要解析 JWT 格式的 access_token，此时需要调用 TokenKeyEndpoint 提供的"/oauth/token_key"端口，获取 JWT 签名时所用的 Key。

（5）WhitelabelApprovalEndpoint 提供的"/oauth/confirm_access"端口，负责构成批准页返回给浏览器。

（6）WhitelabelErrorEndpoint 类提供的"/oauth/error"端口，负责构成错误信息页返回给浏览器。

3. AuthorizationEndpoint 类

提供了两个端口：

（1）GET 类型的"/oauth/authorize"端口，用于接收来自浏览器的授权码请求。

（2）POST 类型的"/oauth/authorize"端口，用户在批准页面选择是否批准的操作后，将进入此端口。

具体源码与解读如下。

```
// 主要判断请求用户是否已经被用户授权
// 如果已授权，则返回新的 authorization_code；反之跳转到用户授权页面
```

```java
@RequestMapping(value = "/oauth/authorize")
public ModelAndView authorize(Map<String, Object> model, @RequestParam
Map<String, String> parameters,
    SessionStatus sessionStatus, Principal principal) {

  // 根据请求参数封装认证请求对象 AuthorizationRequest
  AuthorizationRequest authorizationRequest =
getOAuth2RequestFactory().createAuthorizationRequest(parameters);

  // 获得 response_type 并检验
  Set<String> responseTypes = authorizationRequest.getResponseTypes();

  if (!responseTypes.contains("token") && !responseTypes.contains("code")) {
     throw new UnsupportedResponseTypeException("Unsupported response types: " + responseTypes);
  }

  if (authorizationRequest.getClientId() == null) {
     throw new InvalidClientException("A client id must be provided");
  }

  try {

    if (!(principal instanceof Authentication) || !((Authentication) principal).isAuthenticated()) {
       throw new InsufficientAuthenticationException(
           "User must be authenticated with Spring Security before authorization can be completed.");
    }
    // 获取 OAuth 客户端详细信息
    ClientDetails client =
getClientDetailsService().loadClientByClientId(authorizationRequest.getClientId());

    String redirectUriParameter =
authorizationRequest.getRequestParameters().get(OAuth2Utils.REDIRECT_URI);
    // 如果配置了 client 的 redirect_url，则请求 redirect URL。注意，其必须与 client
    // 配置中的 redirect_url 属性相匹配
    String resolvedRedirect =
redirectResolver.resolveRedirect(redirectUriParameter, client);
       if (!StringUtils.hasText(resolvedRedirect)) {
         throw new RedirectMismatchException(
             "A redirectUri must be either supplied or preconfigured in the
```

```
ClientDetails");
        }
        authorizationRequest.setRedirectUri(resolvedRedirect);

        // 根据 ClientDetail 校验请求的 scope
        oauth2RequestValidator.validateScope(authorizationRequest, client);

        // 需要用到 userApprovalHandler---- > TokenStoreUserApprovalHandler
        authorizationRequest = 
userApprovalHandler.checkForPreApproval(authorizationRequest,(Authentication) 
principal);

        boolean approved = userApprovalHandler.isApproved(authorizationRequest, 
(Authentication) principal);
        authorizationRequest.setApproved(approved);

        // 若已经授权,则直接返回对应的视图。返回的视图中应包含新生成的 authorization_code
        if (authorizationRequest.isApproved()) {
            if (responseTypes.contains("token")) {
                return getImplicitGrantResponse(authorizationRequest);
            }
            if (responseTypes.contains("code")) {
                return new 
ModelAndView(getAuthorizationCodeResponse(authorizationRequest,
                    (Authentication) principal));
            }
        }

        // 将 auth 请求放入 model 中,以便将其存储在会话中供 approveOrDeny 使用
        model.put("authorizationRequest", authorizationRequest);

        // 若未授权用户跳转到授权界面,则由用户自己决定是否授权
        return getUserApprovalPageResponse(model, authorizationRequest, 
(Authentication) principal);

    }
    catch (RuntimeException e) {
        sessionStatus.setComplete();
        throw e;
    }
}
// 处理用户授权页面的结果
```

```java
// 用户是否授予第三方 OAuth 客户端相应的权限
@RequestMapping(value = "/oauth/authorize", method = RequestMethod.POST, params
= OAuth2Utils.USER_OAUTH_APPROVAL)
public View approveOrDeny(@RequestParam Map<String, String> approvalParameters,
Map<String, ?> model,
        SessionStatus sessionStatus, Principal principal) {

    if (!(principal instanceof Authentication)) {
        sessionStatus.setComplete();
        throw new InsufficientAuthenticationException(
                "User must be authenticated with Spring Security before authorizing an access token.");
    }

    // 获取之前存放在 session 中的 authorizationRequest 对象
    AuthorizationRequest authorizationRequest = (AuthorizationRequest)
model.get("authorizationRequest");

    if (authorizationRequest == null) {
        sessionStatus.setComplete();
        throw new InvalidRequestException("Cannot approve uninitialized authorization request.");
    }

    try {
        Set<String> responseTypes = authorizationRequest.getResponseTypes();

        authorizationRequest.setApprovalParameters(approvalParameters);
        // 根据用户是否授权更新 authorizationRequest 对象中的 approved 属性;授予为 true
        authorizationRequest = userApprovalHandler.updateAfterApproval
(authorizationRequest,(Authentication) principal);

        boolean approved = userApprovalHandler.isApproved(authorizationRequest,
(Authentication) principal);
        authorizationRequest.setApproved(approved);

        if (authorizationRequest.getRedirectUri() == null) {
            sessionStatus.setComplete();
            throw new InvalidRequestException("Cannot approve request when no
redirect URI is provided.");
        }

        if (!authorizationRequest.isApproved()) {
```

```java
            return new RedirectView(getUnsuccessfulRedirect(authorizationRequest,
                    new UserDeniedAuthorizationException("User denied access"),
responseTypes.contains("token")),
                    false, true, false);
        }

        if (responseTypes.contains("token")) {
            return getImplicitGrantResponse(authorizationRequest).getView();
        }
        // 生成 code 并存储，以重定向的方式返给 OAuth 客户端
        return getAuthorizationCodeResponse(authorizationRequest,
(Authentication) principal);
    }
    finally {
        sessionStatus.setComplete();
    }
}
```

4. TokenEndpoint 类

TokenEndpoint 类提供了两个端口：

◎ GET 类型的"/oauth/token"端口。

◎ POST 类型的"/oauth/token"端口。

用核心 code 交换 access_token 的功能在 POST 类型的"/oauth/token"端口中，具体源码如下。

```java
// 此方法的主要作用如下:
// 1.获取客户端详情，根据请求参数组装 TokenRequest
// 2.校验请求的 scope
// 3.为客户端生成 token
@RequestMapping(value = "/oauth/token", method=RequestMethod.POST)
public ResponseEntity<OAuth2AccessToken> postAccessToken(Principal principal,
@RequestParam
Map<String, String> parameters) throws HttpRequestMethodNotSupportedException
{
    if (!(principal instanceof Authentication)) {
        throw new InsufficientAuthenticationException(
                "There is no client authentication. Try adding an appropriate
```

```java
authentication filter.");
    }

    // 通过clientId获取OAuth客户端的详细信息
    String clientId = getClientId(principal);
    ClientDetails authenticatedClient =
getClientDetailsService().loadClientByClientId(clientId);

    // 获取请求的相关参数,如grant_type、client_id、scope等(此处会依据 用户的权限进行
    //过滤),封装组建TokenRequest
    TokenRequest tokenRequest =
getOAuth2RequestFactory().createTokenRequest(parameters,
authenticatedClient);

    if (clientId != null && !clientId.equals("")) {
        // 只有当客户端在此请求期间经过身份验证后,才验证客户端的详细信息
        if (!clientId.equals(tokenRequest.getClientId())) {
            // 再次检查,以确保token请求中的客户端ID与经过身份验证的客户端ID相同
            throw new InvalidClientException("Given client ID does not match authenticated client");
        }
    }
    // 对客户端传入的scope进行校验
    if (authenticatedClient != null) {
        oAuth2RequestValidator.validateScope(tokenRequest,
authenticatedClient);
    }
    if (!StringUtils.hasText(tokenRequest.getGrantType())) {
        throw new InvalidRequestException("Missing grant type");
    }
    if (tokenRequest.getGrantType().equals("implicit")) {
        throw new InvalidGrantException("Implicit grant type not supported from token endpoint");
    }
    // 当grant_type=authorzation_code时,清空scope
    if (isAuthCodeRequest(parameters)) {
        // scope是在授权步骤中请求并确定的
        if (!tokenRequest.getScope().isEmpty()) {
            logger.debug("Clearing scope of incoming token request");
            tokenRequest.setScope(Collections.<String> emptySet());
        }
    }
    // 当grant_type=refresh_token时,需要设置scope属性
```

```
    // refresh_token 在 Spring Secuirty OAuth 中被作为一种授权流程看待
    if (isRefreshTokenRequest(parameters)) {
tokenRequest.setScope(OAuth2Utils.parseParameterList(parameters.get(OAuth2Ut
ils.SCOPE)));
    }

    // 验证客户端的 grant_type,并为客户端生成 access_token
    OAuth2AccessToken token =
getTokenGranter().grant(tokenRequest.getGrantType(), tokenRequest);
    if (token == null) {
        throw new UnsupportedGrantTypeException("Unsupported grant type: " +
tokenRequest.getGrantType());
    }
    return getResponse(token);
}
```

14.7.2 资源服务器核心源码分析

资源服务器的核心是 OAuth2AuthenticationProcessingFilter 过滤器,它被插到配置为资源端口的过滤器链中,主要功能是获取请求中携带的 access_token,通过 access_token 提取 OAuth2Authentication 并存入 Spring Security 上下文。

OAuth2AuthenticationProcessingFilter

核心源码如下。

```
public void doFilter(ServletRequest req, ServletResponse res, FilterChain chain)
throws IOException,
        ServletException {
    final boolean debug = logger.isDebugEnabled();
    final HttpServletRequest request = (HttpServletRequest) req;
    final HttpServletResponse response = (HttpServletResponse) res;

    try {
        // 提取请求携带的 token,构建一个认证 Authentication 对象
        Authentication authentication = tokenExtractor.extract(request);

        if (authentication == null) {
            ...
        }
        else {
```

```
request.setAttribute(OAuth2AuthenticationDetails.ACCESS_TOKEN_VALUE,
authentication.getPrincipal());
        if (authentication instanceof AbstractAuthenticationToken) {
            AbstractAuthenticationToken needsDetails =
(AbstractAuthenticationToken) authentication;

needsDetails.setDetails(authenticationDetailsSource.buildDetails(request));
        }
        // 获取token 携带的认证信息。OAuth2AuthenticationMananger 主要做三件事:
        // 1.通过token 获取用户的OAuth2Authentcation 对象(TokenServices)
        // 2.验证访问的资源resourceId是否符合范围
        // 3.验证客户端访问的scope(clientDetailsService)
        Authentication authResult =
authenticationManager.authenticate(authentication);

        ...

        // 将当前的Authentication 放入Context 中,访问后面资源
        SecurityContextHolder.getContext().setAuthentication(authResult);

    }
}
catch (OAuth2Exception failed) {
    ...
}

chain.doFilter(request, response);
}
```

- ◎ TokenExtractor 的默认实现类是 BearerTokenExtractor。
- ◎ AuthenticationManager 的默认实现类是 OAuth2AuthenticationManager。

1. TokenExtractor

此接口的功能是提取请求中包含的 access_token，目前只有一个实现类：BearerTokenExtractor，它只用于提取 Bearer 类型的 access_token。请求中携带的 access_token 参数既可以放在 HTTP 请求头中，也可以在 HTTP 请求参数中。

核心源码如下。

```
protected String extractToken(HttpServletRequest request) {
    // 首先从 header 中解析 access_token
    String token = extractHeaderToken(request);

    // 然后从 request parameter 中解析 access_token
    if (token == null) {
        logger.debug("Token not found in headers. Trying request parameters.");
        token = request.getParameter(OAuth2AccessToken.ACCESS_TOKEN);
        if (token == null) {
            logger.debug("Token not found in request parameters.  Not an OAuth2 request.");
        }
        else {
            request.setAttribute(OAuth2AuthenticationDetails.ACCESS_TOKEN_TYPE, OAuth2AccessToken.BEARER_TYPE);
        }
    }

    return token;
}
```

2. OAuth2AuthenticationManager

OAuth2AuthenticationManager 专门用于处理 OAuth 的资源请求，OAuth2Authentication 对象正是由它负责构建的。由于 OAuth2AuthenticationManager 并没有被添加到 Spring 的 IoC 容器中，所以 Spring Security 管理的普通请求由其他默认的 AuthenticationManager 处理。

核心源码如下。

```
public Authentication authenticate(Authentication authentication) throws AuthenticationException {

    ...
    String token = (String) authentication.getPrincipal();
    //借助 tokenServices，根据 token 加载身份信息
    OAuth2Authentication auth = tokenServices.loadAuthentication(token);
    ...

    checkClientDetails(auth);

    if (authentication.getDetails() instanceof OAuth2AuthenticationDetails) {
        OAuth2AuthenticationDetails details = (OAuth2AuthenticationDetails) authentication.getDetails();
```

```
    ...
}
auth.setDetails(authentication.getDetails());
auth.setAuthenticated(true);
return auth;
}
```

其中，最关键的 tokenSerivces 是 ResourceServerTokenServices 的实例。ResourceServerTokenServices 接口的最主要的两个实现类是 RemoteTokenServices 和 DefaultTokenServices。